Springer Theses

Recognizing Outstanding Ph.D. Research

Aims and Scope

The series "Springer Theses" brings together a selection of the very best Ph.D. theses from around the world and across the physical sciences. Nominated and endorsed by two recognized specialists, each published volume has been selected for its scientific excellence and the high impact of its contents for the pertinent field of research. For greater accessibility to non-specialists, the published versions include an extended introduction, as well as a foreword by the student's supervisor explaining the special relevance of the work for the field. As a whole, the series will provide a valuable resource both for newcomers to the research fields described, and for other scientists seeking detailed background information on special questions. Finally, it provides an accredited documentation of the valuable contributions made by today's younger generation of scientists.

Theses are accepted into the series by invited nomination only and must fulfill all of the following criteria

- They must be written in good English.
- The topic should fall within the confines of Chemistry, Physics, Earth Sciences, Engineering and related interdisciplinary fields such as Materials, Nanoscience, Chemical Engineering, Complex Systems and Biophysics.
- The work reported in the thesis must represent a significant scientific advance.
- If the thesis includes previously published material, permission to reproduce this must be gained from the respective copyright holder.
- They must have been examined and passed during the 12 months prior to nomination.
- Each thesis should include a foreword by the supervisor outlining the significance of its content.
- The theses should have a clearly defined structure including an introduction accessible to scientists not expert in that particular field.

More information about this series at http://www.springer.com/series/8790

Jie Zhang

The Developments and the Applications of the Numerical Algorithms in Simulating the Incompressible Magnetohydrodynamics with Complex Boundaries and Free Surfaces

Doctoral Thesis accepted by
the University of Chinese Academy of Sciences,
Huairou, China

Springer

Author
Dr. Jie Zhang
Chinese Academy of Sciences
Beijing
China

Supervisor
Prof. Mingjiu Ni
Chinese Academy of Sciences
Beijing
China

ISSN 2190-5053 ISSN 2190-5061 (electronic)
Springer Theses
ISBN 978-981-10-6339-8 ISBN 978-981-10-6340-4 (eBook)
https://doi.org/10.1007/978-981-10-6340-4

Library of Congress Control Number: 2018940418

Printed on acid-free paper

This Springer imprint is published by the registered company Springer Nature Singapore Pte Ltd. part of Springer Nature
The registered company address is: 152 Beach Road, #21-01/04 Gateway East, Singapore 189721, Singapore

To my father,
for you endless support on my study
over the last thirty years

Supervisor's Foreword

Magnetohydrodynamics (MHD), which describe the flow behaviors of electrically conducting fluids exposed to the external magnetic fields, have aroused more and more concerns from researchers nowadays. Although it is just a subfield of fluid mechanics, however, it is of great significance in designing some of the key components in the thermonuclear fusion devices, in which the magnetic fields are used to confine the hot plasma. Another respect related to the application of the MHD is the metallurgic industry, which usually employs the magnetic field to control the continuous casting process within the casting mold. Therefore, accurate prediction of the flow behaviors, especially in the multiphase flows, under the influence of the magnetic fields has become a key basic science issue in the fusion and metallurgic engineering. Due to multidisciplinary and multiphysics coupling, it is very difficult to carry out theoretical research work, and thus, experimental studies and numerical methods are excessively relied on to solve the practical engineering problems. Moreover, due to the opacity of the liquid metal, the measuring techniques are rather limited at present to obtain comprehensive information about the flow field. For instance, capturing the vortex structures is rather important in investigating the influence of the magnetic field on the single bubble motion in the liquid metal; however, they are unavailable under the current experimental conditions. Consequently, numerical simulations would be an effective remedy to capture more elaborate structures inside the flow field, whether the fluid is opaque or not.

It should be noted that it is not a simple task to develop a suitable numerical method for the single-phase MHD flows, let alone for the multiphase MHD flows, and the difficulties will be introduced in more detail in this thesis. This Ph.D. thesis by Dr. Jie Zhang has developed a numerical solver aiming to solve the single-phase MHD flows with complex boundaries and the multiphase MHD flows with free surfaces, as being introduced and validated, the algorithms are proved to be accurate and robust by comparing with available analytical and experimental results. It is appropriate and also of important significance that the thought of the engineering science is well inherited in this work, and this solver is applied to simulate some typical interfacial flows probably encountered in the fusion devices, such as the thin liquid lithium film flows, the bubbly flows, and the liquid jet flows. For me, it is a powerful solver to help us in gaining more knowledge in the research community of multiphase MHD flows.

Besides, Dr. Jie Zhang is also skilled in analyzing physical problems involving complicated effects. After completing the coding of the MHD solver, he devoted himself to the numerical research of the bubble motion under the influence of external magnetic fields, since there are corresponding experimental studies on this topic while the results are insufficiently comprehensive because of the opacity of the liquid metal. He has the capability to identify the main or controlling factor in complex situations, to establish the relations between different parameters, and then to discuss the corresponding physical mechanism by analyzing the numerical results. Therefore, one major part of the contents studied in this thesis, i.e., the bubble motion under the influence of either a horizontal or vertical magnetic field, clarifies some interesting phenomena observed but not reasonably interpreted in the experimental studies. In solving these difficulties, Dr. Zhang has demonstrated his broad vision and flexible thinking. He provides plausible explanations on the evolutions of the bubble dynamics when exposed to the external magnetic field, such as why the rising paths of the bubble evolve with different strengths of the magnetic field, and how the vortex structures are modified under the MHD effect. Besides, the anisotropic effect on the flow field induced by the horizontal magnetic field is presented and discussed in this thesis, enriching our knowledge of the MHD effect on the fluid flows although parts of them are predicted by traditional theories.

In addition, after getting more knowledge on the topic of the bubble dynamics, Dr. Zhang tries to answer more general questions in this research community, one of them is why the single bubble always transits from zigzag to spiral, which is another contribution of this thesis. He ascribes this transition to the asymmetric deformation of the bubble shape in the low *Re* regime, and hence, the double-threaded vortex structures behind the bubble will lose symmetry later, leading an angular velocity between them, and finally, the spiral motion is generated. This result, on the one hand, significantly deepens our understanding of the developments of the vortex structures during the bubble rising, and on the other hand, could provide us with the possible way to trigger the bubble rising path transiting from one stage to another.

This thesis has been highly praised by the eight Ph.D. thesis reviewers, and as supervisor of Dr. Zhang, I am glad to recommend this thesis to readers, particularly those specialized or interested in the bubble dynamics and the MHD flows.

Beijing, China
January 2018

Mingjiu Ni

Acknowledgements

First of all, I would like to appreciate the time period of my Ph.D. career from 2009 to 2014, during which time I devoted myself to fluid mechanics from the very beginning to know it, then to research into it and eventually to understand it. It has ever been the most difficult heart-wrenching time, and also the happiest time, over the past years for me. More importantly, it also thought me to keep an average heart with repeated delays and frustrations in research, they were not accidents but certainties.

I would foremost like to express my sincere gratitude toward my supervisor, Prof. Mingjiu Ni, for his continued guidance and generous support throughout my Ph.D. endeavors. Nearly 8 years ago, it was his very presence to bring me into the research community of the fluid mechanics and the magnetohydrodynamics, more than that, it was still him to provide expert advice and encouragement which led me through my Ph.D. career. I am impressed by his experience in academic scholarship and his visions in the academic frontier, and they are unremittingly passed to me during his daily education work. Owing to the 5 years imperceptible influence, both my scientific literacy and ideological level have been greatly promoted. What I am trying to say is, without Prof. Ni, I probably missed the entrance into the hall of sciences.

This thesis could not have reached the same quality without cooperating with professor, also academician of the French Academy of Sciences, René Moreau during his short stay in Beijing. Although I just spent one month with him, however, it was rather a particular period for me that what I learnt from him included, but not limited to, doing research and writing papers. Academician Moreau, in spite of his 75 years ripe old age at that time, still taught me how to analyze the numerical results in detail and how to present them properly in the manuscript. He not only discussed the big-picture ideas with me but also carefully weighed every word and text layout, and even helped to correct punctuation errors many times. Such a spirit and style is worthy to be forever remembered and learned by us younger generation. I wish academician Moreau and his wife good health and longevity!

Second, I am thankful to some other seniors in our research group: Prof. Nianmei Zhang, Prof. Xingang Yu, and Prof. Zenghui Wang; they gave me a lot of help and supports, not only in the academic research but also in my everyday life. I have also benefited a lot from discussions with Prof. Yongling Yu and Prof. Huiyang Ma, who are real experts in fluid mechanics, particularly in the research field of the vortex dynamics; they provide me with the inspiration for the interactions between the bubble motion and the vortex evolutions. I am also grateful to Prof. Stéphane Popinet for sharing the Gerris code he developed, and his dedicated support.

Next, I want to say thanks to brothers and sisters in our research group, it is you who accompany with me throughout my Ph.D. career, and the memories will always remain, forever, present in my heart. Besides, I am also very thankful to some of my friends during the past years; they are Dr. Baiqi Liu, Dr. Qi Xiao, Dr. Chao Yu, Dr. Bing Wu, and Dr. Dong Shao, we have been classmates for more than 10 years since we graduated from the high school.

In addition, I also thank my mother and my elder brother for their unwavering love. Their encouragement strengthened my faith and confidence. On a more personal level, I would like to express my gratitude to my wife, He Zhang for her understanding and support on the way pursuing my passion.

At last, this thesis is dedicated to my father, who gave me constant support and endless love in his short life. It is him partly lead me to this career, and also, I will persist in this way trying to live up to his expectation of myself.

This work was financially supported by the National Natural Science Foundation of China (Grant #11502193, #51636009 and #11125212), the Ministry of Science and Technology of China (Grant #2013GB114000) and the China Postdoctoral Science Foundation (Grant #2015M580834).

Contents

Acronyms

AMR	Adaptive Mesh Refinement
CCM	Cut-cell Method
CFD	Computational Fluid Dynamics
CSF	Continuous Surface Force
CSS	Continuous Surface Stress
DNS	Direct Numerical Simulations
FT	Front Tracking
GCM	Ghost-cell Method
IBM	Immersed Boundary Method
ITER	International Thermonuclear Experimental Reactor
LS	Level-set
LSA	Linear Stability Analysis
MF	Magnetic Field
MHD	Magnetohydrodynamics
PIV	Particle Image Velocity
PLIC	Piecewise Line Interface Calculation
Tokamak	toriodal'naya kamera magnitnymi katushkami
UDV	Ultrasound Doppler Velocimetry
VOF	Volume of Fluid

Chapter 1
Introduction

1.1 Background

1.1.1 Motivation

Throughout the present thesis, magnetohydrodynamics (MHD, hereafter) deal with flows of liquid metal which is subjected to the external magnetic fields (MFs, hereafter), and therefore, the fluid flow is treated as incompressible. MHD is relatively young in natural science and engineering starting with the pioneering work of Hartmann [34] in liquid metal duct flow under the influence of a strong external MF. Under such circumstances, the motion of the electrically conducting fluids generate the electric current and the Lorentz force, and as a consequence, the flow field is coupled and varied with the induced electromagnetic field. Today the research in MHD maybe subdivided into two purposes, there are on the one hand fluid mechanics and applied mathematics because of its complex flow characteristics, such as the MHD instability and the MHD turbulence, which contain particular dynamic behaviors due to the coupling of the flow field and the electromagnetic field. On the other hand, it is also very important in industrial production processes and technology applications, whose working principle is based on the MHD effects, which may be effective in optimizing and controlling the liquid flows in metallurgical and casting industries. Furthermore, the development of MHD pumping devices and the electromagnetic voltmeter also depends on the research in MHD flows. In addition, MHD has been advanced significantly during the last three decades by efforts to take the massive energy out of magnetic-confinement fusion device, which is thought as the most potential energy in future, and therefore, the flow characteristics and the heat transfer in liquid metal flows such as lithium or lithium lead under the influence of MHD effect deserves more attention. The design and construction of a liquid metal fusion blanket or fusion divertor require detailed knowledge of MHD duct flows and MHD free surface flows.

J. Zhang, *The Developments and the Applications of the Numerical Algorithms in Simulating the Incompressible Magnetohydrodynamics with Complex Boundaries and Free Surfaces*, Springer Theses, https://doi.org/10.1007/978-981-10-6340-4_1

1.1.2 The Applications of the MHD Research in the Thermonuclear Fusion Reactor

If light nuclei are forced together, they will fuse with a yield of energy because the mass of the combination will be less than the sum of the masses of the individual nuclei. If the combined nuclear mass is less than that of iron at the peak of the binding energy curve, then the nuclear particles will be more tightly bound than they were in the lighter nuclei, and that decrease in mass comes off in the form of energy according to the Einstein relationship. For elements heavier than iron, fission will yield energy. Fusion is the process that powers active or "main sequence" stars, such as in the sun, however, they belong to the uncontrolled form of thermonuclear fusion, in which the produced energy is released in an uncontrolled manner, as it is also in the thermonuclear weapons such as the "hydrogen bomb". Nevertheless, there is another form of thermonuclear fusion that human beings have long tried to realise: controlled, where the fusion reactions take place in an environment allowing some of the resulting energy to be harnessed for constructive purposes. The thermonuclear fusion reactor aims to attain the goal of "controlled reaction". Nevertheless, the sun uses gravity to get its nuclei close enough and hot enough to start fusion. On Earth, researchers are trying to build fusion reactors of their own. They try to get atoms close enough by maximizing the number of ions in a small region and the amount of time that they stay close together. In order to do that, reactors get heated to temperatures much hotter than the core of the sun ($>10^9$ °C) which converts hydrogen gas into a hydrogen plasma. One choice is to use strong magnetic fields to confine the plasma into a small controllable region can happen, and such magnetic confinement devices are called "Tokamak", which is a transliteration of the Russian words, an acronym of either: ***to**riodal'naya*, ***ka**mera*, ***ma**gnitnymi* and ***ka**tushkami*. Literally, a toroidal magnetic field (traveling around the torus in circles) and a poloidal magnetic field (traveling in circles orthogonal to the toroidal field) are added in tokamak, and the toroidal field is produced by electromagnets surrounding the torus; the poloidal field is the result of a toroidal electric current that flows inside the plasma. This current is induced inside the plasma with a second set of electromagnets. Figure 1.1 is the diagrammatic sketch of the tokamak device, and The heart of a tokamak is its doughnut-shaped vacuum chamber, which contains the charged particles of the plasma being shaped and controlled by the massive magnetic coils placed around the vessel.

For the hot plasma contained by tokamak, the deuterium-tritium fusion reaction seems the most likely path, and the schematic drawing is presented in Fig. 1.2. It shows the fusion of deuterium with tritium creates helium-4, freeing a neutron, and releasing 17.59 MeV as kinetic energy of the products while a corresponding amount of mass disappears, in agreement with kinetic $E = \Delta mc^2$, where Δm is the decrease in the total rest mass of particles. However, very large kinetic energy is needed to overcome the repulsive electrostatic force, which is also called "Coulomb barrier", with the aid of tunneling, requiring very high temperatures. In addition, sufficient

Fig. 1.1 The diagrammatic sketch of the tokamak device

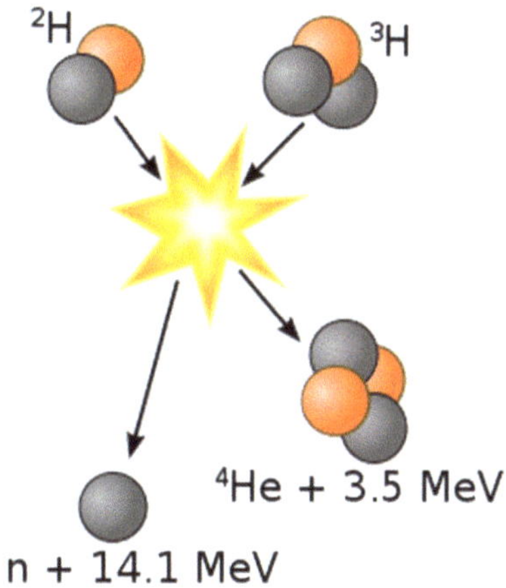

Fig. 1.2 The schematic diagram for deuterium - tritium fusion reaction, 80% of that energy yield is in the energy of the neutron

density and energy confinement are also required to trigger the thermonuclear fusion reaction, and that is why external magnetic fields are applied in the reactor since no solid material could withstand the extremely high temperature of the plasma.

Moreover, ITER (formerly "International Thermonuclear Experimental Reactor") is an international nuclear fusion research and engineering megaproject, which will be the world's largest tokamak nuclear fusion reactor that is building in southern France. The ITER project aims to make the long-awaited transition from experimental studies of plasma physics to full-scale electricity-producing fusion power stations. The ITER fusion reactor has been designed to produce 500 megawatts of output power for around twenty minutes while needing 50 megawatts to operate. Thereby the machine aims to demonstrate the principle of producing more energy from the fusion process than is used to initiate it, something that has not yet been achieved in any fusion reactor. Moreover, the project also has some other aims, such as to maintain a fusion pulse for up to 8 min, to ignite a "burning" (self-sustaining) plasma, and to verify tritium breeding concepts. Regarding the technical design, ITER contains several key components for energy production and conversion: the vacuum vessel, the breeder blanket, the magnet system, the additional heating system, the cryostat and also the cooling system.

When considering the MHD effect in the tokamak fusion device, the liquid metal flows in the blanket (Fig. 1.3a) that shields the steel vacuum vessel and external machine components from high-energy neutrons produced during the fusion reaction are greatly influence. The function of the blanket is twofold: first of all, the neutrons produced by the fusion reactions - which are not confined by the magnetic fields as they have no charge - penetrate the blanket and react with lithium, producing the fusion fuel tritium. The tritium can then be extracted, processed and added to deuterium for refuelling the reactor. On the other hand, the blanket is used to remove the heat deposited by the fast neutrons. The heat is removed by a cooling fluid, and in a commercial reactor can be used to produce electricity. Therefore, the liquid metal flows inside the blanket are greatly influenced by the MFs, particularly the electromagnetic resistance on the flow, and the heat transfer efficiency is considerably decreased. Due to the difficulty in obtaining such strong MFs (11 Tesla) experientially, numerical simulations would be an effective method to study the MHD flows in the blankets. Besides, due to the complex structures of the blanket, it is important to develop an efficient numerical technique to simulate complex boundary bounded MHD flows. Furthermore, the liquid-gas MHD flows may be also encountered for the purpose of heat transfer enhancement through injecting bubble plumes into the blanket, as a consequence, the numerical tools are required to simulate the multiphase MHD flows.

Other components that suffer from the MHD effect are under high heat load such as the plasma facing first wall or plasma divertors and limiters (Fig. 1.3b), which allow the online removal of waste material from the plasma while the reactor is still operating. This allows control over the buildup of fusion products in the fuel, and removes impurities in the plasma that have entered into it from the vessel lining. For a solid divertor, it may be difficult to provide adequate armour lifetime, armour-heat sink joint lifetime (CfC-Cu and W-Cu) and heat sink lifetime, sustaining thermo-hydraulic and electro-mechanical loads. Therefore, the free surface flows of the liquid lithium on the divertor would be an alternative option. Under such circumstance, the capillary force, the electromagnetic force, the inertial force and the viscous force are almost the same order of magnitude. Consequently, it would be important to study the spatial-temporal evolutions of the free surface of the liquid lithium under the influence of strong MFs, and the numerical simulations will undoubtedly help a lot.

(a) (b)

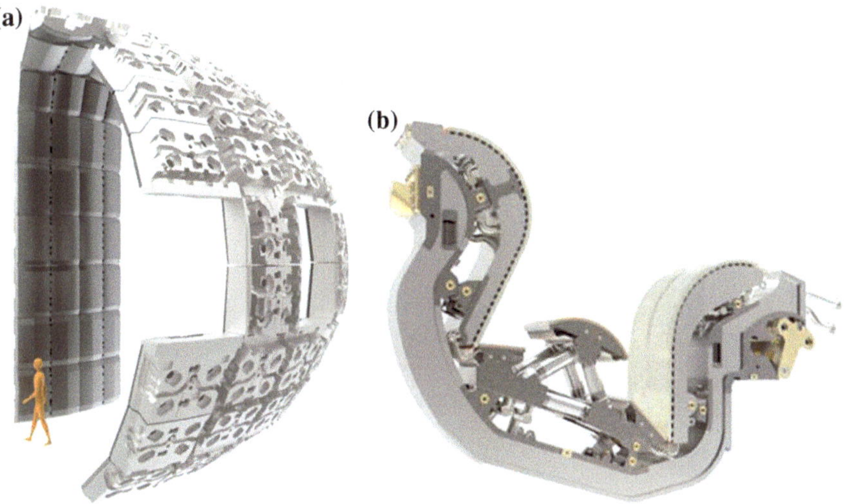

Fig. 1.3 Sketch of different components in the tokamak fusion device that suffer from the MHD effect. **a** The banket; **b** the divertor

1.1.3 The Applications of the MHD Research in the Metallurgy Industry

Besides the applications in the thermonuclear fusion reactor, for many processes in chemical and metallurgic engineering MHD effect is also worth investigation. For instance, high current density processing is employed for heating and melting metallic materials in induction furnaces, in which the heat is applied by induction heating of metal [13, 43]. In Fig. 1.4a, the induction furnace is depicted, it consists of a nonconductive crucible holding the charge of metal to be melted, surrounded by a coil of copper wire. The MF induces eddy currents, circular electric currents, inside the metal, by electromagnetic induction [24]. The eddy currents, flowing through the electrical resistance of the bulk metal, heat it by Joule heating. There is additionally a preferred mixing effect within the melt by electromagnetic stirring due to induced Lorentz forces. The movement within the melt is indicated in Fig. 1.4a by instantaneous streamline patterns.

On the other hand, a rising field of the MHD application in the metallurgic industry is the control of the continuous casting process by a combined electromagnetic braking and moderate stirring in the inlet region of the casting mold [62]. With the help of the electromagnetic stirring of liquid metal within the mold, the quality of the strand surface/sub-surface and the solidification structure, i.e. structure refinement, soundness and chemical homogeneity, can be improved. This is schematically shown in Fig. 1.4b. It should be noticed that in the continuous casting of steel, the Argon gas bubbles are sometimes injected through the submerged entry nozzle to prevent it from clogging, or to homogenize the physical and chemical properties of the melt

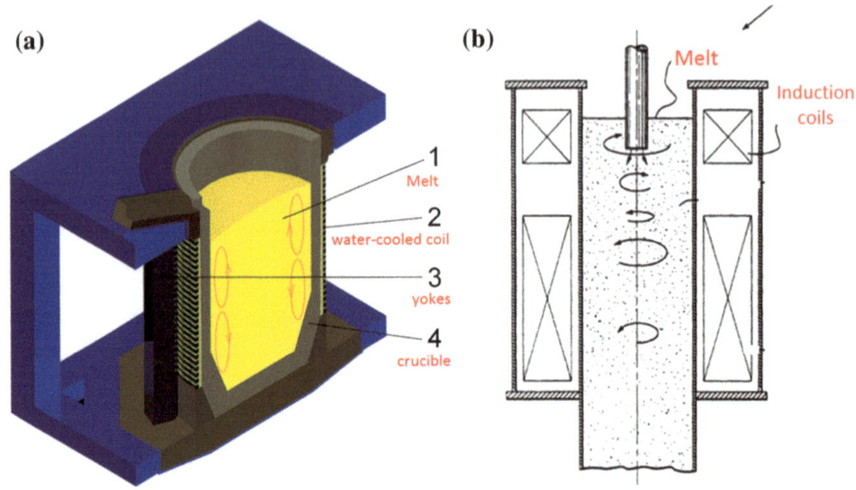

Fig. 1.4 Sketch of **a** induction furnace; **b** continuous casting

[1, 40, 105]. Therefore, the MF should have a considerable influence on the bubble velocity, the bubble shape or the bubble distribution of gas in the cross section of the mold [52, 114]. Although the electrically insulating gas bubbles do not experience a direct impact of the electromagnetic force, however, the pressure and the velocity field in the surrounding conducting fluid are strongly affected by the applied MF.

Another actual field of the MHD control in industrial engineering is the monocrystal pulling process that a moderately imposed MF may be the means to increase the size of the pulled crystals while maintaining the quality. Herein, the surface tension driven capillary flow is dominated in the melt pool, moreover the temperature field is, in most instance, inhomogeneous at the free surface, and therefore the Marangoni effect is significant to induce the thermocapillary flow. Such type of MHD flows with inhomogeneous surface tension also require detailed research [38, 78, 109].

1.2 The Development of the Numerical Algorithms for Simulating MHD Flows

Since the present thesis can be subdivided into two parts, respectively of the development of the numerical algorithms in simulating the MHD flows and the physics of the single bubble motion under the influence of the external MF, therefore, the following literature reviews are also presented separately.

1.2.1 For Single Phase MHD Flows with Complex Boundaries

Two-dimensional (2D) MHD channel flows have been extensively studied by theoretical analysises since Hunt [39], Shercliff [85] and walker [104], who gave the distributions of the induced electromagnetic field when the liquid metal fluids in an electrically conducting or insulating conduct were subjected to a transverse MF. Since then, such analytical solutions provided good benchmarks for the validations of the numerical algorithms developed for simulating the MHD flows.

With the accelerative growth of high - performance computers, and also owing to the application of more advanced numerical schemes in computational fluid dynamics (CFD, hereafter), numerical simulations give rise to more opportunity in studying the MHD flows, especially of those three-dimensional (3D) MHD phenomena. However, difficulties are always encountered with regard to the much thinner boundary layer(typically $\delta \sim Ha^{-1}$, Ha is the Hartmann number describing the ratio between the electromagnetic force and the viscous force) compared with the hydrodynamic flows (typically $\delta \sim Re^{-1/2}$, Re is the Reynolds number describing the ratio between the inertial force and the viscous force), consequently the 3D direct numerical simulations (DNS, hereafter) require very thin grids to fully capture the boundary layer in vicinity of the walls. On the other hand, the lack of good numerical schemes for calculating the electromagnetic fields also limits the development of DNS in simulating high Ha MHD flows, which can be as high as $Ha = 10^4 \sim 10^5$ in the fusion blankets. As the author know, to develop such an accurate and efficient numerical algorithm has long been a problem and hence the 3D numerical simulations of inertial flows are often limited to the steady regime and low Ha.

For low magnetic Reynolds numbers (Re_m describing the relative strength between the induced MF and the external MF), the electrical potential formula can be employed for MHD with good accuracy [56, 62]. By using a staggered grid system to avoid the checkerboard phenomenon of pressure, as shown in Fig. 1.5a, Sterl [88] conducted the DNS of liquid metal MHD flows in the rectangular conducts by using the uniform Cartesian grids, on which the current density was easily conserved. However, within such a uniform spatial resolution, as we mentioned before, it would be a great challenge of the computational resources if the boundary layers were fully resolved at high Ha flows. In his computations, Ha was restricted to a maximum value of 10^3 in the 2D simulations while it decreased to 50 in 3D. However, regarding the actual MHD flows inside the blankets of the tokamak devices, Ha is always higher than 10^4 and therefore the nonuniform grids are required for practical utilization. Nevertheless, for the non-uniform grids with large scale ratio between the main flow region and the boundary layer in high Ha flows, typically on the scale of 10^3 or larger, it brings new challenges in designing a robust numerical scheme to calculate the current density and the Lorentz forces. By using a non-uniform staggered grid system, Leboucher [45] found the classic second-order scheme for electrical potential gradient could not lead to an accurate solution of the current density inside the boundary layer. After recognizing this, he then designed a fully staggered grid

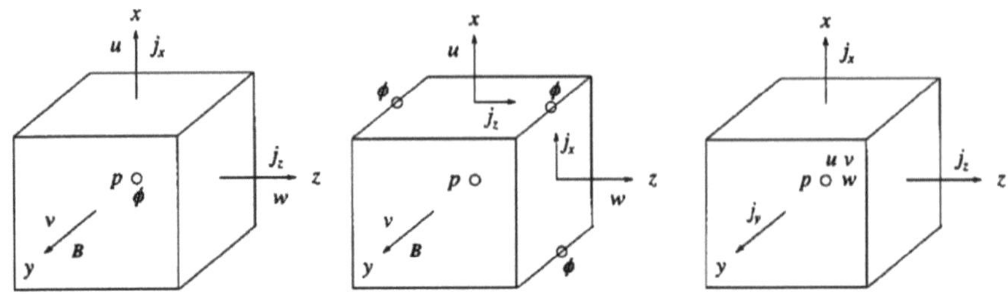

(a) Staggered grids (b) Fully staggered grids (c) Collocated grids

Fig. 1.5 Three grid arrangement systems

system, which required the pressure to be located at the grid center, the velocities at the center of the grid faces, and the electric potential at the center of the grid lines, as shown in Fig. 1.5b. Moreover, the locations of the current flux were more complex being depended on the direction of the MF. By using such a complicated grid system, accurate results could be obtained in simulating high Ha flows when the MF was unidirectional. Afterwards, the fully staggered grids were also used for large eddies simulation [87]. However obviously, the spatial discretisations of the numerical scheme would be much more complicated if the geometry of the flow domain was complex or if the MFs were multidirectional, as a consequence, its practical utilization was also limited.

More recently, Ni et al. [65] proposed a consistent and conservative scheme for simulating MHD flows based on the collocated grids, on which all the variables were located at the center of the grid, as shown in Fig. 1.5c. This numerical algorithm not only simplified the spatial dicretisations but also maintained the conservative property of the current density in solving the electromagnetic field. In addition, they identified that the keys in designing this scheme were in two aspects: there was on one hand to construct a consistent scheme to calculate the electric potential across the grid face and on the other hand to construct another consistent scheme to calculate the Lorentz force. The numerical solutions obtained by their algorithms were in good agreement with the available analytical solutions and the experimental results, even if the Hartmann number is as high as $Ha \sim 10^4$ and the aspect ratio of the grid scale was also larger than 10^3. Further then, Ni et al. [66, 67] extended this numerical method to the unstructured grids and the staggered grids.

Besides, the geometrical structures of the blankets in the fusion device are more complex than the rectangular ducts, and therefore, the numerical simulations of the MHD flows within such complex boundaries lead to higher requirement for grid generations. Traditionally, there are two types of grids to handle such problem: the body-fitted grid and the unstructured grid, as shown in Fig. 1.6, where the flow past an airfoil is simulated within the two grids in (a) and (b). In both grids, the grid faces conform to the geometry of the flow domain, and hence the boundary conditions are imposed directly on the grid faces. Regarding their application in simulating the MHD flows, Ni et al. [66] used the unstructured grids to study the pipe flows under

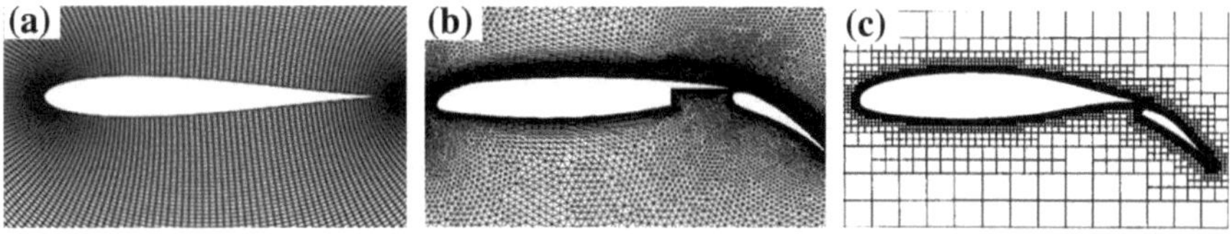

Fig. 1.6 Different grid strategies in simulating flows past the airfoil. **a** Body-fitted grids; **b** unstructured grids; **c** Cartesian grids

the influence of a transverse MF, and the numerical solutions agreed very well with the experimental results. Furthermore, based on the same grids, Vantieghem et al. [102] also investigated the influence of the electrically conducting wall on the MHD pipe flows. In addition, Mutschke et al. [63] used the body-fitted grids to study the MHD flows past a circular cylinder, indicating that the three dimensional instability were not always suppressed by the streamwise MF. Even though both grids prove to be applicable and accurate in simulating the MHD flows bounded by complex boundaries, however as widely know, it is rather time consuming in grid generation and coordinate transforming.

More recently, a new grid generation technique, known as the "Cartesian grids" to be appeared in Fig. 1.6c, is likely to draw more attention in the CFD research community not only because of its automatic generation character but also because of its more adaptive performance when the geometries of the boundary are complex. During the last two decades, three types of Cartesian grids are developed and applied in sequence: the immersed boundary method (IBM), the ghost cell method (GCM) and the cut-cell method (CCM). IBM was firstly developed by Peskin [71] to simulate cardiac mechanics and associated blood flow that the immersed boundaries on grids did not conform to the geometry of the heart, and alternatively, the effect of the immersed boundary on the flow was viewed as a source term in solving the Navier–Stokes equations. The main advantage of such a technique is to free the computers from coordinate transformation or complex discretization operators that it significantly reduce the per-grid-point operation count. Afterwards, due to the noticeable advantages, numerous efforts have been devoted to developing this method, which have been successfully used for a wide variety of problems including cardiac mechanics [72], cochlear dynamics [5] and bubble dynamics [101]. However, IBM is a kind of "continuous method" that the immersed boundary is artificially constructed with finite thickness and the smoothing of the force distribution probably leads to an inability to provide a sharp representation of the IB and this can be especially undesirable for high *Re* flows. Therefore, within such flows, the new Cartesian grid technique with "sharp interface" is required that the boundary conditions are imposed directly on the IB with no numerical diffusion. GCM is one approach falling into this category that the boundary conditions are enforced through constructing the "ghost cells" inside the solid body. Different interpolation schemes are developed [50, 98] to calculate the velocities or pressure on the ghost cells, in order to satisfy the boundary conditions. This method has been used to simulate a wide variety of flows including compressible flow past a circular cylinder and an airfoil [29] at Reynolds numbers up to $O(10^5)$, and aquatic propulsion [55]. However, GCM as a kind a finite-difference approach, which suffers from the non-conservative property that the strict conservation law of both mass and momentum can not be fully guaranteed. Correspondingly, that is why CCM being further introduced because of its advantage in conserving the mass and momentum of the fluid, viewed as a finite-volume approach. Literally, this method allows the Cartesian grids cut by the solid interface to be identified and reshaped after the intersection of the boundary with the sides of these cut cells is determined, and next, those cells lying in the solids are discarded. Furthermore, as a requirement of the finite-volume discretization, appropriate interpolating schemes are needed to estimate the pressure gradients, the convective flux and the diffusive flux on the the partial cell faces, as developed in variant forms [54, 99, 100, 110]. It should be noted that within the CCM, the reshaped cells cut by the solid boundaries may be rather small that only a minor portion is retained in the fluid region, in this case, the computational timestep is more strict and hence the merging technique for the cut cells is also developed [74, 77].

Although the Cartesian grids have been widely used in simulating the ordinary fluid flows, nevertheless, only the IBM discussed so far has been applied to simulate the MHD flows by Grigoriadis et al. [31]. They added a continuous source term in the discrete form of the electromagnetic equations, in order to enforce the electrically insulating boundary condition on the solid wall. Through this, they managed to simulate the MHD flows inside a circular pipe within the Cartesian grids and the numerical results were in good agreement with the theoretic solutions. However, the weakness of this approach is also obvious that it will be invalid when the solid wall is electrically conducting, because there is no definite boundary condition for the current density any longer along the fluid-solid boundaries. By contraries, the CCM is able to overcome such difficulty by split the mixed cell into two parts: partially immersed in the fluid and partially in the solid, and then the two parts can be treated numerically with a sharp scheme by reconstructing the fluid-solid interface.

1.2.2 For Multi-phase MHD Flows

As introduced in Sects. 1.1.2 and 1.1.3, the bubbly driven circulation systems are used in the blanket of the fusion device as well as in the metal processing operations such as steel making and secondary refining of metal, and therefore, to understand the dynamic behaviors of the gas-liquid multiphase flows under the influence of external MF is not only scientific but also engineering importance. However, due to the nonlinear properties of the coupling Maxwell and Navier–Stokes equations, and also because of the difficulty in tracking the positions of the gas-liquid interface on a real-time level, exact analytic solutions for such problems are really rare. In addition, the experimental studies on the multi-phase MHD flows are not easy either. For instance, the liquid lithium, which is of practical interest in the tokamak device, is almost impossible for the optical access to the flow field because the liquid metal is opaque. Although there are some new measuring techniques being developed, such as the ultrasound Doppler velocimetry (UDV) method [113, 114] and the electric potential probe method [57], however, they are still far from fully revealing the flow field, particularly the evolution of the vortex structures and the bubble shapes, which are rather important for identifying the flow characteristics. For the reasons given above, the numerical simulations become an alternative choice for the multi-phase MHD research community. Different from the single-phase flows, the numerical schemes in simulating the multi-phase flows need to track the translating and deforming interface, which seems to be a moving boundary problem. Regarding the corresponding numerical approaches, they can be divided into two categories generally, that is the "one-fluid" method where a singe set of governing equations is used to describe the fluid motion in both fluids, and in the another approach, the grid faces fully conform with the fluid-fluid interface and the governing equations are solved in different fluids separately. Because the former is much more widely used due to their simplicity, some representative approaches belonging to it are introduced herein.

- *The Volume-of-Fluid method*

The Volume-of-Fluid (VOF, hereafter) method was first discussed by Hirt and Nichols [36] and was then developed to be a more accurate and "sharp" approach. Within this method, the interface is "reconstructed" in such a way that the interface will not flow into a new grid until the current grid is full, and f is used to represent the fraction of phase 1 in a given cell. As a consequence, $0 < f < 1$ indicates the existence of the interface in the cell, while $f = 0$ or $f = 1$ means the grid is "empty" or "full" of phase 1, respectively. To reconstruct the interface, a series of different numerical methods were proposed: Noh and Woodward [69] gave a simple line interface calculation (SLIC) method, Youngs [108] further introduced a more complex method of piecewise line interface calculation (PLIC), which reconstructed the interface with a line (2D) or a plane (3D) that could be oriented arbitrarily in such a way as to best fit the interface. Since then, more promising methods are developed, such as the least-squares algorithms proposed by Pilliod and Puckett [73], the cubic spline interpolation method given by López et al. [46], and the improved PLIC scheme constructed by Gueyffier et al. [32].

After reconstructing the interface, the volume fraction of phase 1 will be advanced by solving a convection equation, by which the mass is fully conserved. A particular direction-split scheme is always used to advance the interface, as firstly proposed by DeBar [18] and further improved by Rider and Kothe [80], Rudman [81], R. Scardovelli and S. Zaleski [83]. Besides, to decrease the artificial thickness of the interface, the reconstructed geometric interface is further used to calculate the flux across the interface as a sharp approach.

- *The Level-Set method*

The Level-Set (LS) method was introduced by Osher and Sethian [70], but its first application in tracking the fluid-fluid interface belonged to the pioneering work of M. Sussman et al. [92] and Chang et al. [14] who used it to simulate the bubble motion and the droplets falling problems. This approach uses a continuous level-set function to identify the position of the interface as the zero contour of the level-set function. Therefore, a smooth transition zone is generated across the "physical" interface that the "numerical" interface has thickness larger than one grid space. This property increase the regularity of the interface in comparison with the VOF method. However, it also has a big disadvantage that it can not conserve the mass very well while transporting the interface. Further then, many researchers focus to improve the conservative property of the Level-Set method, such as the Level-Set-VOF combination scheme proposed by M. Sussman and Puckett [93] and the technique to decrease spurious flows [35].

- *The Front-Tracking method*

The Front-Tracking (FT) method was introduced by Unverdi and Tryggavason [101] for multi-phase flows where the interface was marked by connected marker points. It should be noted that only the calculations of the markers are done on the Lagrangian grids while the rest of the computations are completed on a fixed Eulerian grid system, as similar as what be done in the VOF method. Therefore, nowadays, methods using marker points are generally referred to as front-tracking methods to distinguish them from front-capturing methods were the marker function is advected directly. One of the main objections to the use of FT methods is that topology changes, where fluid regions merge or breakup, are not handled automatically. Changing the connection of the front points can obviously be accomplished, but at the cost of increased code complexity. In FT, the interfaces never fuse together unless something special is done and sometimes the added level of control provided by the tracking is desirable. The question of how interfaces merge is still not a fully resolved issue.

Moreover, as aforementioned, the calculation of the surface tension force is another difficulty in numerically simulating the interfacial flows. From a physical point of view, the surface tension only acts along the free surface to prevent the interface from deforming, therefore, an idea implementation is to impose the surface tension force concentrating at the interface as local point source. However, such singularity will introduce numerical difficulty because the discontinuous force will lead to great local gradient that the numerical instabilities or nonphysical oscillations may happen. To solve this problem, Brackbill et al. [8] introduced a Continuous Surface Force (CSF) model to distribute the surface tension force into the neighbor grids in vicinity to the interface, thereby there was a "numerical thickness" of the interface. Similarly, Lafaurie et al. [42] also developed a Continuous Surface Stress (CSS) model as a variant method to solve this singularity problem. Other models include some smoothing of the theoretically ideally sharp interface to enable calculation of gradients and are suitable for the implementation as volumetric sources within the finite-volume framework, rather than discontinuities. Nevertheless, such treatments suffer another problem, which is described as the "spurious flow" that the artificial velocities may lead to non-physical instabilities close to the interface if the separated fluids are of high density ratio or large surface tension. Until recently Renardy and Renardy [79] and Francois et al. [25] noticed that the calculation of the pressure gradient and the surface tension across the interface should be balanced to decrease the spurious flow, and this method was acknowledged as the "balanced force" model. Moreover, Ni [68] further validated that the key point in the numerical treatment was to use a consistent algorithm to discrete the pressure gradient and surface tension force, by which the two forces were always balanced even if they were not calculated synchronously.

Another difficulty in tackling the surface tension force is how to calculate the interface curvature within the VOF framework, because inaccurate approach would also leads to spurious flows. By quantitative calculations, Cummins [16] validated that the height function method was more accurate than either the convolved VOF technique or the reconstructed distance function. In addition, by comparing with other highly accurate numerical algorithms, such as the parabola fitting approach [79] or the curve fitting approach [30], height function is thought to be more simple to implement.

In another circumstance, the temperature distribution may vary along the liquid-gas interface, and therefore the surface tension gradients are produced to induce tangential stresses, known as Marangoni effects to drive the fluid flowing in the vicinity of the interface. A typical problem where thermal Marangoni stresses play a significant role is the thermocapillary migration of drops and bubbles, much of the available work in this research community has been reviewed by Subramanian and his co-workers [89]. The analytical study of the thermocapillary motion of drops and bubbles could be found in the pioneering work of Young et al. [107] although their solution was only valid in a very limited parameter spaces, namely zero Reynolds number and Marangoni number. After that, a series of theoretical work were conducted to consider the convective effect in a wider parameter space [2, 3, 90, 91]. In addition, the first 3D numerical simulation of the thermocapillary motion of deforming drops was presented by Haj-Hariri et al. [33] by using the LS method to capture the deforming interface. Nas and Tryggvason [64] used the FT method to simulate the thermocapillary motion of a pair of 3D fully deformable bubbles and drops. More recently, Ma and Bothe [48] developed a new numerical approach to simulate the thermocapillary flows within the VOF framework that particular treatment was required to calculate the gradient of the surface tension.

Regarding the multi-phase MHD flows, the publications about the developments of the numerical algorithms are scarcely available to our knowledge. A fully 3D numerical scheme to simulate the incompressible free surface MHD flows has been implemented in HIMAG code [58] by using the LS method for interface tracking. R. Samulyak et al. [82] simulated multi-phase MHD flows based on a hyperbolic system by using a FT technique to capture the interface, in which only the discretization of the electrical potential Poisson equation is described in details that they could not obtain accurate results when the external MF was strong. Takatani [94] also proposed a mathematical model to simulate the multi-phase MHD flows that a marker and cell method was employed for solving the Navier–Stokes equations while the VOF method was used to track the interface. However, their numerical algorithms were not presented in details either for the calculation of the Lorentz force or for the estimation of the surface tension force, and the validations were not given yet. Therefore, not only the analytical solutions for the multi-phase MHD flows, but also the numerical simulations for such problems are really rare, and they increase the difficulties in this research community.

1.3 The Rising Bubble in the Viscous Liquids

Bubbly flows are widely used in the metals processing, as aforementioned to stir or to refine the melt, and in such an aggressive environment, MFs, either induced by the process itself as in induction furnaces or additionally applied as in continuous casters, are one of the rare potential tools to control the bubble motion. Besides, in the fusion devices, where liquid lithium or lithium-lead are used as a breeder or as a heat remover, it has been proposed [58] to employ the type of the two-phase flows to reduce the average electrical conductivity and so the head losses, without reducing much neither the global heat transfer nor the breeding effect. Those examples demonstrate that the motion of either a single bubble or bubble plumes in a liquid metal in the presence of MFs is a topic of great interest.

1.3.1 Without Magnetic Fields

It is known nowadays that the bubble shape, the rising path and the trailing wake do not develop independently during the bubble motion, which is rather characterized by the interactions between different influential factors. Leonardo Da Vinci was the first one to report this idea [15, 76], denoted as the Leonardos paradox by the scientific community. Although there are numerous researches being carried out in the past half a century, irrespectively with the theoretical, experimental and numerical studies, however, the transition of the bubble motion from the rectilinear path to the unstable path is not fully understood, and the physical mechanisms causing such transition are still under investigation.

By using hyper-purified water, Duineveld [20] confirmed for the first time that the bubble rising in water would present oscillatory path when the bubble radius $R > 0.91$ mm, corresponding to a Reynolds number of $Re \approx 660$. Thereafter, more and more experiments were carried out to study the relationships between the wake structures and the path instabilities. Lunde and Perkins [47] used the Dye Visualizations to observe the vortex structures, while Ellingsen and Risso [23] used the Laser Doppler Anemometry and Brücker [9] employed the Particle Image Velocity (PIV) method. According to their observations, the spirally rising path was observed to associate with a continuous wake composed of double-threaded vortex pairs, while the zigzag motion was accompanied with the shedding of hairpin vorticities. However, most of the experiments were suffered from contamination because of the additives mixed in the liquid for experimental measurements, under such circumstance, the bubble interface was not shear-free any longer and it was more likely to be no-slip. Other techniques introduce temperature gradients over the bubble surface, inducing surface tension gradients and consequently, secondary currents caused by the Marangoni effect. Then in order to keep the liquids be highly purified during the experiments, significant efforts were made over the last fifteen years [19, 103, 112], and most of them observed a shedding process of double-threaded vortices at the rear of the bubble during the zigzag stage, while the vortex structures were more continuous and stable in the spiral stage. Moreover, by using different silicon oils, which did not need an ultrapure environment, Zenit and Magnaudet [112] found that it was actually the aspect ratio χ, a dimensionless measure of the shape deformation, rather than Re to trigger the path instability.

Because of the difficulty in developing a suitable visualizing technique in highly purified water, numerical simulations were gradually employed to study the bubble dynamics by keeping the bubble with a frozen shape. Mougin and Magnaudet [59] proved that the path instability was coupled with the unstable wake, they imposed manual perturbations to cause the bubble motion to transit from zigzag to spiral. After that, Mougin and Magnaudet [60] and Shew and Pinton [86] both found the evolutions of the forces and torques acting on the bubble were different when it traveled within a zigzag and a spiral path respectively. Magnaudet and Mougin [49] reported the critical value of Re and χ for the appearance of wake instability by studying the uniform flow past an frozen ellipsoidal bubble, as an extension of the numerical investigations by G. Leal [44] and Blanco and Magnaudet [6]. On basis of this, Cano-Lozano et al. [10] performed a more precise study with realistic fore-and-aft asymmetric bubble shapes instead of the ellipsoidal bubble.

On the other hand, as far as the direct numerical simulations are concerned to allow the bubble to deform freely under gravity, the available studies are really rare to the author's knowledge. Gaudlitz [26] and Hua [37] were able to show periodic vortex shedding during the zigzag motion, and the later work identified different vortex structures during the zigzag and spiral motions, the results were fully consistent with the experimental observations. Within the VOF framework, Cano-Lozano et al. [11] and Tripathi et al. [97] confirmed that the double-threaded vortices were shed periodically in the zigzag stage while the vortex structures were more stable in the spiral stage. It should be noted that by performing more than twenty numerical cases with refined grids, Cano-Lozano et al. [11] gave a phase diagram summarizing different patterns of the rising paths observed in their simulations, and the transition border between different stages agreed well with the experimental results. That figure identifies the parameter spaces within which the bubble does not follow a vertical, straight path, and eventually, it give the transition condition when the zigzag bubble enters the spiral stage.

Another important approach in this research community is the Linear Stability Analysis (LSA), which is used to determine the transition curve. Tchoufag et al. [95] find the neutral curve by considering the flow past a frozen bubble whose orientation and shape are fixed. Later on, they consider the influence of the freedom degree by coupling the linearized governing equations between the fluid and the freely rising bubble [96]. Alternatively, Cano-Lozano et al. [12] study the flows past the bubble with a fore-aft shape instead of the spheroid shape, and more accurate results are observed.

1.3.2 With Magnetic Fields

When the single bubble rises in the liquid metal, the experimental studies are more complex due to some extra difficulties. On one hand, liquid metals are prone to be oxidized, and in general a melt is never free of contaminants so that an oxide layer forms at the gas-liquid interface, and therefore, the velocity at the bubble interface is more like to be no-slip rather than free shear stress. On the other hand, most of the liquid metals are opaque and consequently, the above mentioned experimental measurement technology for ordinary liquids, such as the Dye Visualization and the PIV methods are not applicable any more. Instead, Ultrasound Doppler Velocimetry (UDV) method is an alternative approach under such circumstance and has been used to study the motion of bubbly flows [113, 114] under the influence of MFs. Local conductivity probes have also been used to measure the rise velocity of bubbles in mercury [57] as well as the behavior of gas bubbles in turbulent liquid metal MHD flows [21, 22]. Both difficulties explained the scarcity of the published experimental data, however the limited available results still shed some light on this topic, they revealed that the effect of the MF on the single bubble motion was quite subtle that small bubble was accelerated by moderate MF wile larger one was obstructed. Another relevant experiment was conducted by G. Gerbeth et al. [28] about a spherical droplet falling through an inhomogeneous horizontal MF. They found the droplet to be elongated in the direction of MF, and the shape oscillations were observed when it approached the exit of the nonuniform MF.

Besides the numerical difficulties in simulating the multi-phase flows, as described in Sect. 1.2.2, the direct numerical simulations of the bubble motion in liquid metal under the influence of MF are even more challenging due to other respects: the differences of the physical properties between the two fluids are more significant, and the computation of the surface tension and the Lorentz force are coupled. As a consequence, there are very few phase-resolving simulations of bubbles in liquid metal under the influence of a MF. Although some attempts [27, 52] have been made, either by using the business software CFX with a modified turbulence model or with a hybrid particle-level-set method, the quantitative conclusions were not satisfactory. More recently, by applying a IBM method to impose the slippery boundary condition on the bubble interface, Schwarz and Fröhlich [84] were able to simulate the single bubble rising in the liquid GaInSn under the influence of a vertical MF, but they supposed the bubble shape to be frozen during its motion.

It is known that the homogeneous MF substantially modify the vortical structures, no matter in the turbulence flows [7, 41] or in the MHD flows past obstacles [61, 63]. Therefore, since the vorticity dynamics behind the bubble is rather crucial in most of the cases, the impact of the MF on the bubble motion is expected to be significant. In particular, the influence of the MF on the interactions between the bubble wake and the bubble motion deserves deep investigation, and one of the important goals in the present study is to fulfill this gap.

1.4 The Gerris Flow Solver

Throughout this thesis, the MHD module and other modules are developed based on the Gerris software, which is a powerful open source code developed by S. Popinet [74, 75], aiming to resolves the time-dependent incompressible Navier–Stokes equations by using the finite-volume method based on the collocated spatial discretisation. All the variables are located in the center of the grids in order to avoid the extra complexity brought about by the staggered grids, however, particular interpolation schemes are needed to prevent the pressure decoupling problem [53, 111]. An approximate projection method [51] is applied to solve the velocity-pressure coupling equations that a multigrid method [106] is used to accelerate the iterative solution when solving the pressure Poisson equation. Besides, the advection term is discretized with a conservative Godunov scheme, which is developed by Bell, Colella and Glaz [4]. The evaluation of the diffusion term employs a Crank–Nicolson scheme, which is semi-implicit and second-order accurate.

One of the most distinctive features of the Gerris Flow Solver is its quadtree/octree mesh structure, which allows to adapt the simulation mesh to conform to the repartition of the small and large scales structures of the problem of interest, as shown in Fig. 1.7 that the left one describes the meshes are locally refined around the solid circular while the right one indicates the meshes in vicinity of the bubble interface are also refined. By using this adaptive mesh refinement (AMR) technique, resolution is optimised so that only the regions where small scale resolution is needed are refined very finely. It should be also noticed that different criterions can be defined according to the user's research purpose, that is, besides the existing criterions according to the gradient and vorticity magnitude, other criterions can be implemented.

To deal with the embedded solid boundaries, Gerris uses the cut-cell method as depicted in Fig. 1.7a, it is observed that all the grids in the solid region are destroyed

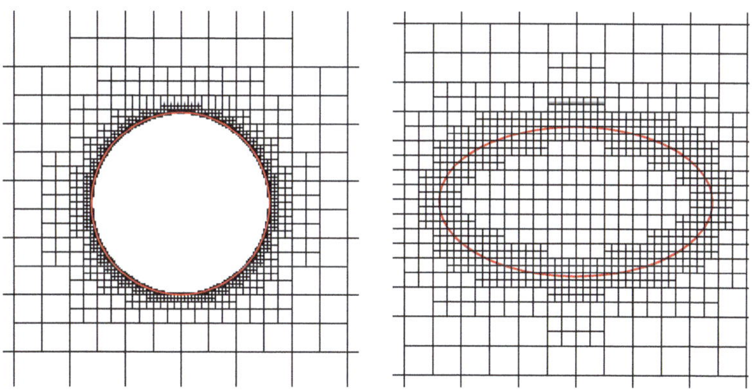

(a) Solid-fluid boundaries (b) Fluid-fluid interface

Fig. 1.7 Details of the spatial resolution during the numerical simulations by using the quad/octree adaptive AMR technique. **a** The meshes near the solid-fluid boundaries; **b** the meshes near the fluid-fluid interface and eliminated. In addition, the grids cut by the solid boundaries are not fully cubic, they are called as mixed-cells in the following study. The characteristics of this method is discussed in Sect. 1.2.1 and more details can be found in Ref. [74], and hence they will not be repeated here.

Regarding the numerical algorithms in simulating the multi-phase flows, Gerris implements a piecewise-linear VOF advection scheme, and the interface is reconstructed with a geometrical manner that the flux across the interface is very sharp. By applying the AMR technique, Gerris allows a dynamic adaptive mesh refinement along the interface, as shown in Fig. 1.7b, and by means of different criterions according to the magnitude of the gradient or the vorticity, the spatial discretisation is adjusted to follow the scale and temporal evolution of flow structures without a significant global increment of computational resources. As demonstrated by Cano-Lozano et al. [10], Gerris ia more advantageous than OpenFoam in simulating the bubbly flows, because the spurious flows around the interface are significantly reduced within the numerical framework of Gerris. This is ascribed to the accurate estimation of the curvature of the interface by implementing a generalized height-function method, and the balance-force continuum formulation [8, 25] also helps stabilize the flows.

On basis of the existing numerical schemes mentioned above, the first part of the present manuscript will show how the single-phase MHD module is implemented, the multi-phase MHD module and the Marongoni force module into Gerris.

1.5 Content and Innovations of This Thesis

1.5.1 Research Contents

Specifically, the work of this thesis is composed of the following aspects:

- **The development of the numerical schemes in simulating the single-phase MHD flows with complex boundaries**. In the first part of Chap. 3, a consistent and conservative scheme for calculating the electromagnetic field is implemented in Gerris, particularly to be adapted in the AMR framework, the scheme is able to keep the consistent property of the current flux across the fine/coarse mesh boundaries. When the geometry of the flow domain is complex and electrically insulating, the Cut-cell method is applied in solving the electromagnetic field on the Cartesian grids, and this is much more efficient than the traditional body-fitted or unstructured grids. In addition, when the wall is electrically conducting, the situation would be more complex because the electromagnetic field needs to be solved in the whole domain while the flow field is only solved in the fluid domain. Under such circumstance, the conjugate calculation of the electromagnetic field needs to identify the solid-fluid boundary first by using a VOF method, and afterwards, a cut-cell approach is employed to solve the electric potential and the current density based on the Cartesian grids. The validations of the numerical methods are presented in Chap. 4.

- **The development of the numerical schemes in simulating the multi-phase MHD flows with Marangoni effects**. By implementing the MHD module in the multi-phase solver of Gerris, as described in the second part of Chap. 3, we find accurate results are obtained even with high ratio ($>10^4$) of electric conductivities between the two fluids. The validations of the code are presented by comparing the results with available theoretical and experimental solutions. This new solver can be further applied to study the influence of the external MF on the multi-phase flows, which are frequently encountered in the fusion devices and the metallurgic engineering, such as the bubble rising and the drop splashing. The validations of the numerical methods are also presented in Chap. 4.
- **The numerical investigation of the single bubble motion under the influence of a vertical or a horizontal MF.** In Chap. 5, we identify the close relationships between the "wake instability" and "path instability" during the bubble motion when path transition happens. In addition, by imposing a constant MF whether in the vertical or horizontal direction, we investigate the its influence on the rising velocity of the bubble, the deformation of the bubble shape, the wake structure behind the bubble and the rising path of the bubble. It is found by changing the bubble shape, the MF is also able to modify the vortex structures and the rising path of the bubble significantly. In addition, bubble motion shows a "second instability" under strong vertical MF while the flow field presents anisotropic characteristics under a horizontal MF. These phenomena can not be observed in the experiments because the liquid metal is opaque, however, with the help of the numerical simulations, we get more comprehensive understanding of the magnetic influence on the bubble motion.
- **The investigation of the physical mechanisms to drive the bubble motion from zigzag to spiral**. In Chap. 7, the physical mechanism to cause the path transition of a rising bubble from zigzag to spiral is investigated numerically. Several groups of direct numerical simulations are performed that equal-strength vortex pairs are observed to shed twice during a period of the pure zigzag path, and this type of motion is triggered by the amounts of streamwise vorticities accumulated on the bubble interface, when a critical value is reached. However, when the balance between the counter-rotating vortices is broken, an angular velocity is induced between the asymmetric vortex pairs, driving the bubble to rise in an opposite spiral path. Besides, as the imbalance between the vortex pairs will be weakened under MF, a vertical MF is imposed to study the effect after the vortex pairs become more symmetric. Finally, we find the rotating radius of the spirally rising bubble is decreased, indicating that our interpretation is credible.

1.5.2 Innovation of This Work

In this thesis, we have developed a numerical solver designed for simulating the imcompressible MHD flows, either with complex boundaries or with free surface. Besides its academic significance, it could be also used in the engineering practice for fusion device or metallurgic industry. We think there are several innovations that are noteworthy in the research community, as shown below:

1. By simulating the MHD flows based on an AMR framework, the computational efficiency would be greatly improved because the scale ratio of the grids lying in the core flow field and boundary layer would be larger than 10^3, the automatic AMR technique avoids generating the non-uniform grids manually. Besides, the approach handles the embedded solid boundary based on the Cartesian grids, which save the computational resources since the generation of the body-fitted grids is really time consuming. Within the same framework, the MHD flows bounded by the electrically conducting walls can be simulated on the "one mesh system".
2. By performing direct numerical simulations, we are able to give more reliable interpretations to the dynamics behaviors of the bubble motion under the influence of a MF. The MHD effect on the bubble shape and on the vortex structures are investigated numerically, and they can not be detected in the experiments because the liquid metal is opaque. Therefore, this part of our work fulfill the gap between the experimental data and the hidden behind physics. More important, we confirm that the horizontal MF will produce an anisotropic flow field in vicinity of the bubble, and such directional effect is in good agreement with the theoretical predictions given by P.A. Davidon [17] and R. Moreau et al. [56].
3. The identification of the physical mechanism to drive the bubble transiting from zigzag to spiral. Previous studies always described the different behaviors of bubble, or different vortex structures behind the bubble. Our mechanisms, which is based on the quantitative calculation of the vortex strengths, however, explains how different vortex structures are evolved, and how an angular velocity is induced due to unsymmetrical vortex pairs to push the bubble traveling from zigzag to spiral. Moreover, by imposing external MF, which is used to diminish the asymmetry between the vortex pairs, this mechanisms is further validated.

References

1. Bai H, Thomas BG (2001) Bubble formation during horizontal gas injection into downward-flowing liquid. Metall Mater Trans B 32(6):1143–1159
2. Balasubramaniam R, Subramaniam RS (1996) Thermocapillary bubble migrationłthermal boundary layers for large Marangoni numbers. Int J Multiph Flow 22(3):593–612
3. Balasubramaniam R, Subramanian RS (2000) The migration of a drop in a uniform temperature gradient at large Marangoni numbers. Phys Fluids 12(4):733–743
4. Bell JB, Colella P, Glaz HM (1989) A second-order projection method for the incompressible Navier-Stokes equations. J Comput Phys 85(2):257–283
5. Beyer RP (1992) A computational model of the cochlea using the immersed boundary method. J Comput Phys 98(1):145–162
6. Blanco A, Magnaudet J (1995) The structure of the axisymmetric high - Reynolds number flow around an ellipsoidal bubble of fixed shape. Phys Fluids 7(6):1265–1274

7. Boeck T, Krasnov D, Zienicke E (2007) Numerical study of turbulent magnetohydrodynamic channel flow. J Fluid Mech 572:179–188
8. Brackbill JU, Kothe DB, Zemach C (1992) A continuum method for modeling surface tension. J Comput Phys 100(2):335–354
9. Brücker C (1999) Structure and dynamics of the wake of bubbles and its relevance for bubble interaction. Phys Fluids 11(7):1781–1796
10. Cano-Lozano JC, Bohorquez P, Martłnez-Bazn C (2013) Wake instability of a fixed axisymmetric bubble of realistic shape. Int J Multiph Flow 51:11–21
11. Cano-Lozano JC, Martinez-Bazan C, Magnaudet J et al (2016) Paths and wakes of deformable nearly spheroidal rising bubbles close to the transition to path instability. Phys Rev Fluids 1(5):053604
12. Cano-Lozano JC, Tchoufag J, Magnaudet J et al (2016) A global stability approach to wake and path instabilities of nearly oblate spheroidal rising bubbles. Phys Fluids 28(1):014102
13. Campbell FC (ed) (2013) Metals fabrication: understanding the basics. ASM International
14. Chang YC, Hou TY, Merriman B et al (1996) A level set formulation of Eulerian interface capturing methods for incompressible fluid flows. J Comput Phys 124(2):449–464
15. Clift R, Grace JR, Weber ME (2005) Bubbles, drops, and particles. Courier Corporation
16. Cummins SJ, Francois MM, Kothe DB (2005) Estimating curvature from volume fractions. Comput Struct 83(6):425–434
17. Davidson PA (1995) Magnetic damping of jets and vortices. J Fluid Mech 299:153–186
18. DeBar R (1974) Fundamentals of the KRAKEN code. Technical report UCIR-760
19. De Vries AWG, Biesheuvel A, Van Wijngaarden L (2002) Notes on the path and wake of a gas bubble rising in pure water. Int J Multiph Flow 28(11):1823–1835
20. Duineveld PC (1995) The rise velocity and shape of bubbles in pure water at high Reynolds number. J Fluid Mech 292:325–332
21. Eckert S, Gerbeth G, Lielausis O (2000) The behaviour of gas bubbles in a turbulent liquid metal magnetohydrodynamic flow: Part I: dispersion in quasi-two-dimensional magnetohydrodynamic turbulence. Int J Multiph Flow 26(1):45–66
22. Eckert S, Gerbeth G, Lielausis O (2000) The behaviour of gas bubbles in a turbulent liquid metal magnetohydrodynamic flow: Part II: magnetic field influence on the slip ratio. Int J Multiph Flow 26(1):67–82
23. Ellingsen K, Risso F (2001) On the rise of an ellipsoidal bubble in water: oscillatory paths and liquid-induced velocity. J Fluid Mech 440:235–268
24. Erickson RW, Maksimovic D (2007) Fundamentals of power electronics. Springer Science and Business Media, Berlin
25. Francois MM, Cummins SJ, Dendy ED et al (2006) A balanced-force algorithm for continuous and sharp interfacial surface tension models within a volume tracking framework. J Comput Phys 213(1):141–173
26. Gaudlitz D, Adams NA (2009) Numerical investigation of rising bubble wake and shape variations. Phys Fluids 21(12):122102
27. Gaudlitz D, Adams NA (2010) The influence of magnetic fields on the rise of gas bubbles in electrically conductive liquids. Direct and large-eddy simulation VII. pp 465–471
28. Gerbeth G, Kaudze M, Gailitis A (1989) Deformation of an electrically conducting drop in a magnetic field. Liq Metal Magnetohydrodyn 10:287
29. Ghias R, Mittal R, Lund TS (2004) A non-body conformal grid method for simulation of compressible flows with complex immersed boundaries. AIAA paper 80(2004):6
30. Ginzburg I, Wittum G (2001) Two-phase flows on interface refined grids modeled with VOF, staggered finite volumes, and spline interpolants. J Comput Phys 166(2):302–335
31. Grigoriadis DGE, Kassinos SC, Votyakov EV (2009) Immersed boundary method for the MHD flows of liquid metals. J Comput Phys 228(3):903–920
32. Gueyffier D, Li J, Nadim A et al (1999) Volume-of-fluid interface tracking with smoothed surface stress methods for three-dimensional flows. J Comput Phys 152(2):423–456
33. Haj-Hariri H, Shi Q, Borhan A (1997) Thermocapillary motion of deformable drops at finite Reynolds and Marangoni numbers. Phys Fluids 9(4):845–855

34. Hartmann J (1937) Hg-Dynamics-I. Math-Fys, MLA
35. Herrmann M (2008) A balanced force refined level set grid method for two-phase flows on unstructured flow solver grids. J Comput Phys 227(4):2674–2706
36. Hirt CW, Nichols BD (1981) Volume of fluid (VOF) method for the dynamics of free boundaries. J Comput Phys 39(1):201–225
37. Hua J, Lin P, Stene JF (2008) Numerical simulation of gas bubbles rising in viscous liquids at high Reynolds number. Moving interface problems and applications in fluid dynamics, contemporary mathematics, vol 466:17–34
38. Huang H, Zhou X (2009) The impact of normal magnetic fields on instability of thermocapillary convection in a two-layer fluid system. J Heat Transf 131(6):062502
39. Hunt JCR (1965) Magnetohydrodynamic flow in rectangular ducts. J Fluid Mech 21(4):577–590
40. Iguchi M, Tokunaga H, Tatemichi H (1997) Bubble and liquid flow characteristics in a woods metal bath stirred by bottom helium gas injection. Metall Mater Trans B 28(6):1053–1061
41. Knaepen B, Moreau R (2008) Magnetohydrodynamic turbulence at low magnetic Reynolds number. Ann Rev Fluid Mech 40:25–45
42. Lafaurie B, Nardone C, Scardovelli R et al (1994) Modelling merging and fragmentation in multiphase flows with SURFER. J Comput Phys 113(1):134–147
43. Laughton MA, Say MG (2013) Electrical engineer's reference book. Elsevier
44. Leal LG (1989) Vorticity transport and wake structure for bluff bodies at finite Reynolds number. Phys Fluids A Fluid Dyn 1(1):124–131
45. Leboucher L (1999) Monotone scheme and boundary conditions for finite volume simulation of magnetohydrodynamic internal flows at high Hartmann number. J Comput Phys 150(1):181–198
46. López J, Hernández J, Gómez P et al (2004) A volume of fluid method based on multidimensional advection and spline interface reconstruction. J Comput Phys 195(2):718–742
47. Lunde K, Perkins RJ (1997) Observations on wakes behind spheroidal bubbles and particles. In: ASME fluids engineering division summer meeting, Vancouver, Canada, p 141
48. Ma C, Bothe D (2011) Direct numerical simulation of thermocapillary flow based on the volume of fluid method. Int J Multiph Flow 37(9):1045–1058
49. Magnaudet J, Mougin G (2007) Wake instability of a fixed spheroidal bubble. J Fluid Mech 572:311–337
50. Majumdar S, Iaccarino G, Durbin P (2001) RANS solvers with adaptive structured boundary non-conforming grids. In: Annual research briefs, Center for Turbulence Research, Stanford University, pp 353–466
51. Martin DF, Colella P (2000) A cell-centered adaptive projection method for the incompressible Euler equations. J Comput Phys 163(2):271–312
52. Miao X, Lucas D, Ren Z et al (2013) Numerical modeling of bubble-driven liquid metal flows with external static magnetic field. Inter J Multiph Flow 48:32–45
53. Minion ML (1996) A projection method for locally refined grids. J Comput Phys 127(1):158–178
54. Mittal R, Bonilla C, Udaykumar HS (2003) Cartesian grid methods for simulating flows with moving boundaries. Computational methods and experimental measurements-XI. pp 557–566
55. Mittal R, Seshadri V, Udaykumar HS (2004) Flutter, tumble and vortex induced autorotation. Theor Comput Fluid Dyn 17(3):165–170
56. Moreau RJ (2013) Magnetohydrodynamics. Springer Science and Business Media, Berlin
57. Mori Y, Hijikata K, Kuriyama I (1977) Experimental study of bubble motion in mercury with and without a magnetic field. J Heat Transf 99(3):404–410
58. Morley NB, Smolentsev S, Munipalli R et al (2004) Progress on the modeling of liquid metal, free surface, MHD flows for fusion liquid walls. Fusion Eng Des 72(1):3–34
59. Mougin G, Magnaudet J (2001) Path instability of a rising bubble. Phys Rev Lett 88(1):014502
60. Mougin G, Magnaudet J (2006) Wake-induced forces and torques on a zigzagging/spiralling bubble. J Fluid Mech 567:185–194

61. Mück B, Günther C, Mller U et al (2000) Three-dimensional MHD flows in rectangular ducts with internal obstacles. J Fluid Mech 418:265–295
62. Müller U, Bühler L (2013) Magnetofluiddynamics in channels and containers. Springer Science and Business Media, Berlin
63. Mutschke G, Gerbeth G, Shatrov V et al (1997) Two-and three-dimensional instabilities of the cylinder wake in an aligned magnetic field. Phys Fluids 9(11):3114–3116 (MLA)
64. Nas S, Tryggvason G (2003) Thermocapillary interaction of two bubbles or drops. Int J Multiph Flow 29(7):1117–1135 (MLA)
65. Ni MJ, Munipalli R, Morley NB et al (2007) A current density conservative scheme for incompressible MHD flows at a low magnetic Reynolds number. Part I: on a rectangular collocated grid system. J Comput Phys 227(1):174–204
66. Ni MJ, Munipalli R, Huang P et al (2007) A current density conservative scheme for incompressible MHD flows at a low magnetic Reynolds number. Part II: on an arbitrary collocated mesh. J Comput Phys 227(1):205–228
67. Ni MJ, Li JF (2012) A consistent and conservative scheme for incompressible MHD flows at a low magnetic Reynolds number. Part III: on a staggered mesh. J Comput Phys 231(2):281–298
68. Ni MJ (2009) Consistent projection methods for variable density incompressible Navier-Stokes equations with continuous surface forces on a rectangular collocated mesh. J Comput Phys 228(18):6938–6956
69. Noh WF, Woodward P (1976) SLIC (simple line interface calculation). In: Proceedings of the fifth international conference on numerical methods in fluid dynamics, 28 June–2 July 1976, Twente University, Enschede. Springer, Berlin, pp 330–340
70. Osher S, Sethian JA (1988) Fronts propagating with curvature-dependent speed: algorithms based on Hamilton-Jacobi formulations. J Comput Phys 79(1):12–49
71. Peskin CS (1972) Flow patterns around heart valves: a numerical method. J Comput Phys 10(2):252–271
72. Peskin CS (1982) The fluid dynamics of heart valves: experimental, theoretical, and computational methods. Ann Rev Fluid Mech 14(1):235–259
73. Pilliod JE, Puckett EG (2004) Second-order accurate volume-of-fluid algorithms for tracking material interfaces. J Comput Phys 199(2):465–502
74. Popinet S (2003) Gerris: a tree-based adaptive solver for the incompressible Euler equations in complex geometries. J Comput Phys 190(2):572–600
75. Popinet S (2009) An accurate adaptive solver for surface-tension-driven interfacial flows. J Comput Phys 228(16):5838–5866
76. Prosperetti A (2004) Bubbles. Phys Fluids 16(6):1852–1865
77. Quirk JJ (1994) An alternative to unstructured grids for computing gas dynamic flows around arbitrarily complex two-dimensional bodies. Comput Fluids 23(1):125–142 (MLA)
78. Rao AR, Biswal PC (2001) Benard-Marangoni convection in two thin immiscible liquid layers with a uniform magnetic field. Acta Mech 151(1):61–73
79. Renardy Y, Renardy M (2002) PROST: a parabolic reconstruction of surface tension for the volume-of-fluid method. J Comput Phys 183(2):400–421
80. Rider WJ, Kothe DB (1998) Reconstructing volume tracking. J Comput Phys 141(2):112–152
81. Rudman M (1997) Volume-tracking methods for interfacial flow calculations. Int J Numer Methods Fluids 24(7):671–691
82. Samulyak R, Du J, Glimm J et al (2007) A numerical algorithm for MHD of free surface flows at low magnetic Reynolds numbers. J Comput Phys 226(2):1532–1549
83. Scardovelli R, Zaleski S (2003) Interface reconstruction with least - square fit and split Eulerian C Lagrangian advection. Int J Numer Methods Fluids 41(3):251–274
84. Schwarz S, Fröhlich J (2014) Numerical study of single bubble motion in liquid metal exposed to a longitudinal magnetic field. Int J Multiph Flow 62:134–151
85. Shercliff JA (1956) The flow of conducting fluids in circular pipes under transverse magnetic fields. J Fluid Mech 1(6):644–666
86. Shew WL, Pinton JF (2006) Dynamical model of bubble path instability. Phys Rev Lett 97(14):144508

87. Shimomura Y (1991) Large eddy simulation of magnetohydrodynamic turbulent channel flows under a uniform magnetic field. Phys Fluids A Fluid Dyn 3(12):3098–3106
88. Sterl A (1990) Numerical simulation of liquid-metal MHD flows in rectangular ducts. J Fluid Mech 216:161–191
89. Subramanian RS, Balasubramaniam R (2001) The motion of bubbles and drops in reduced gravity. Cambridge University Press, Cambridge
90. Subramanian RS (1981) Slow migration of a gas bubble in a thermal gradient. AIChE J 27(4):646–654
91. Subramanian RS (1983) Thermocapillary migration of bubbles and droplets. Adv Space Res 3(5):145–153
92. Sussman M, Smereka P, Osher S (1994) A level set approach for computing solutions to incompressible two-phase flow. J Comput Phys 114(1):146–159
93. Sussman M, Puckett EG (2000) A coupled level set and volume-of-fluid method for computing 3D and axisymmetric incompressible two-phase flows. J Comput Phys 162(2):301–337
94. Takatani K (2007) Mathematical modeling of incompressible MHD flows with free surface. ISIJ Int 47(4):545–551
95. Tchoufag J, Magnaudet J, Fabre D (2013) Linear stability and sensitivity of the flow past a fixed oblate spheroidal bubble. Phys Fluids 25(5):054108
96. Tchoufag J, Magnaudet J, Fabre D (2014) Linear instability of the path of a freely rising spheroidal bubble. J Fluid Mech 751
97. Tripathi MK, Sahu KC, Govindarajan R (2015) Dynamics of an initially spherical bubble rising in quiescent liquid. Nat Commun 6:6268
98. Tseng YH, Ferziger JH (2003) A ghost-cell immersed boundary method for flow in complex geometry. J Comput Phys 192(2):593–623
99. Udaykumar H, Shyy W, Rao M (1996) ELAFINT-A mixed Eulerian–Lagrangian method for fluid flows with complex and moving boundaries. In: 6th Joint thermophysics and heat transfer conference, 1996
100. Udaykumar HS, Mittal R, Shyy W (1999) Computation of solid-liquid phase fronts in the sharp interface limit on fixed grids. J Comput Phys 153(2):535–574
101. Unverdi SO, Tryggvason G (1992) A front-tracking method for viscous, incompressible, multi-fluid flows. J Comput Phys 100(1):25–37
102. Vantieghem S, Albets-Chico X, Knaepen B (2009) The velocity profile of laminar MHD flows in circular conducting pipes. Theor Comput Fluid Dyn 23(6):525–533
103. Veldhuis C, Biesheuvel A, Van Wijngaarden L (2008) Shape oscillations on bubbles rising in clean and in tap water. Phys Fluids 20(4):040705
104. Walker JS (1981) Magneto-hydrodynamic flows in rectangular ducts with thin conducting walls. 1. Constant-area and variable-area ducts with strong uniform magnetic-fields. Journal de Mcanique 20(1):79–112
105. Wang Z, Mukai K, Izu D (1999) Influence of wettability on the behavior of argon bubbles and fluid flow inside the nozzle and mold. ISIJ Int 39(2):154–163
106. Wesseling P (1995) Introduction to multigrid methods
107. Young NO, Goldstein JS, Block MJ (1959) The motion of bubbles in a vertical temperature gradient. J Fluid Mech 6(3):350–356
108. Youngs DL (1982) Time-dependent multi-material flow with large fluid distortion. Numerical methods for fluid dynamics
109. Yang M, Ma N (2005) A computational study of natural convection in a liquid-encapsulated molten semiconductor with a horizontal magnetic field. Int J Heat Fluid Flow 26(5):810–816
110. Ye T, Mittal R, Udaykumar HS et al (1999) An accurate Cartesian grid method for viscous incompressible flows with complex immersed boundaries. J Comput Phys 156(2):209–240
111. Zang Y, Street RL, Koseff JR (1994) A non-staggered grid, fractional step method for time-dependent incompressible Navier–Stokes equations in curvilinear coordinates. J Comput Phys 114(1):18–33

112. Zenit R, Magnaudet J (2008) Path instability of rising spheroidal air bubbles: a shape-controlled process. Phys Fluids 20(6):061702
113. Zhang C, Eckert S, Gerbeth G (2005) Experimental study of single bubble motion in a liquid metal column exposed to a DC magnetic field. Int J Multiph Flow 31(7):824–842
114. Zhang C, Eckert S, Gerbeth G (2007) The flow structure of a bubble-driven liquid-metal jet in a horizontal magnetic field. J Fluid Mech 575:57–82

Chapter 2
Governing Equations

2.1 Basic Equations

The set of equations governing the viscous incompressible fluid motion under the influence of an external MF are the momentum and mass conservation equations, given by

$$\rho\left(\frac{\partial \boldsymbol{u}}{\partial t}+\boldsymbol{u}\cdot\nabla\boldsymbol{u}\right)=-\nabla p+\nabla\cdot\mathbb{S}+\boldsymbol{F}_s+\boldsymbol{F}_l+\boldsymbol{S} \tag{2.1}$$

$$\nabla\cdot\boldsymbol{u}=0 \tag{2.2}$$

where $\mathbb{S}$ is the viscous stress tensor written as

$$\mathbb{S}=2\mu\mathbb{D}=2\mu\left(\frac{1}{2}(\nabla\boldsymbol{u}+\nabla\boldsymbol{u}^T)\right)$$

with density ρ, pressure p and dynamics viscosity μ. $\boldsymbol{F}_s$ stands for the surface tension which just acts on the interface and is formulated as a volume force based on the CSF technique. With a constant surface tension coefficient, it has a formulation of:

$$\boldsymbol{F}_s=\sigma\kappa\delta_s\boldsymbol{n} \tag{2.3}$$

where σ is the surface tension coefficient, κ the interface curvature and $\boldsymbol{n}$ the normal direction of interface, the force confined on interface is represented by Dirac distribution function δ_s. In addition, $\boldsymbol{F}_l$ is the induced Lorentz force and it takes the form of

$$\boldsymbol{F}_l=\boldsymbol{J}\times\boldsymbol{B} \tag{2.4}$$

J. Zhang, *The Developments and the Applications of the Numerical Algorithms in Simulating the Incompressible Magnetohydrodynamics with Complex Boundaries and Free Surfaces*, Springer Theses, https://doi.org/10.1007/978-981-10-6340-4_2

where $\boldsymbol{J}$ is the induced current density and $\boldsymbol{B}$ is the external MF. The volumetric electromagnetic forces have a conservation form [2] when the applied magnetic field is constant. It is given as

$$\boldsymbol{F}_l = \nabla \cdot \mathbb{T}_l = \nabla \cdot (\boldsymbol{J}(\boldsymbol{r} \times \boldsymbol{B})) \tag{2.5}$$

where $\mathbb{T}_l$ is the electromagnetic stress tensor, $\boldsymbol{r}$ is a distance vector.

The last term at the right hand side of Eq. 2.1 indicates other source terms contributing to the fluid flow such as gravity.

When the Marangoni effect due to inhomogeneous surface tension coefficient is considered, the calculation of the temperature field is required. Therefore, the energy conservation equation is coupled with Eq. 2.1 and Eq. 2.2 in the form of

$$\frac{\partial T}{\partial t} + \boldsymbol{u} \cdot \nabla T = \nabla \cdot \left(\frac{\lambda}{\rho c_p} \nabla T \right) \tag{2.6}$$

where c_p denotes the heat capacity and λ the heat conductivity. It should be noted that the viscous dissipating term and the Joul heating term are neglected because they are comparatively small in general.

Correspondingly, because the surface tension coefficient is now depended on the temperature gradient, and hence the additional tangential Marangoni stresses should be calculated, and Eq. 2.3 is renewed with

$$\boldsymbol{F}_s = \sigma \kappa \delta_s \boldsymbol{n} + \nabla_{\parallel}(\sigma)\delta_s \tag{2.7}$$

where the first term on the righthand side corresponds to the normal stress balance and $\nabla_{\parallel}$ in the second term indicates the thermocapillary force being tangent to the interface. Furthermore, σ is suggested to vary linearly with temperature as

$$\sigma = \sigma_0 + \sigma_T(T - T_0) \tag{2.8}$$

which is an accurate approximation for most fluids. In the expression, σ_0 is the surface tension coefficient at the reference temperature of T_0 while σ_T represents the change rate which is assumed to be negative with temperature for most fluids of interest. Then Eq. 2.7 has the form of

$$\boldsymbol{F}_s = \sigma \kappa \delta_s \boldsymbol{n} + \sigma_T \nabla_{\parallel}(T)\delta_s \tag{2.9}$$

Afterwards, let us return to the solution of the Lorentz forces in Eq. 2.4. There are two numerical approaches to evaluate the current density [1]: either by coupling between the Ohm's law and the Maxwell's equations to solve the induced MF, or by solving the electric potential equation obtained from the Ohm's law. In the present work, the second method is employed because the induced MF is negligible in comparison with the external MF that the magnetic Reynolds number, which has the form of $R_m = \mu^* \sigma_e L_0 u_0$, is far less than unity for liquid metal flows in a small scale.

In the formulation, μ^* is the magnetic permeability, σ_e is the electric conductivity, while L_0 and u_0 are the character scales of the length and the velocity, respectively.

For fluids with velocity of $\boldsymbol{u}$, Ohm's law defines the current density in the form of

$$\boldsymbol{J} = \sigma_e(\boldsymbol{E} + \boldsymbol{u} \times \boldsymbol{B}) \tag{2.10}$$

where $\boldsymbol{E}$ is the induced electric field. When the induced MF is much smaller than the external field, this equation can be written as

$$\boldsymbol{J} = \sigma_e(-\nabla\varphi + \boldsymbol{u} \times \boldsymbol{B}) \tag{2.11}$$

where φ is the induced electric potential. According to the charge conservation law, we must enhance

$$\nabla \cdot \boldsymbol{J} = 0 \tag{2.12}$$

therefore a combined equation is derived from Eqs. 2.11 and 2.12, given as the so called electric potential Poisson equation:

$$\nabla \cdot (\sigma_e \nabla\varphi) = \nabla \cdot (\sigma_e(\boldsymbol{u} \times \boldsymbol{B})) \tag{2.13}$$

2.2 Dimensionless Numbers

It should be noticed that some dimensional variables are always introduced to rescale the Navier-Stokes equations. If the characteristic length is L_0, the characteristic velocity is u_0, the temperature gradient in the bulk fluid is G_T, and the characteristic magnetic intensity is B_0, then the variant dimensionless variables are given as:

$$L^* = \frac{L}{L_0},\ \boldsymbol{u}^* = \frac{\boldsymbol{u}}{u_0},\ t^* = \frac{tu_0}{L_0},\ p^* = \frac{p}{\rho u_0^2},$$
$$\kappa^* = L_0\kappa,\ \boldsymbol{B}^* = \frac{\boldsymbol{B}}{B_0},\ \boldsymbol{j} = \frac{\boldsymbol{j}}{\sigma_e u_0 B_0},\ \varphi_0 = \frac{\varphi}{L_0 u_0 B_0},\ T^* = \frac{T}{G_T L_0}$$

Correspondingly, the governing equations of Eqs. 2.1 and 2.2 are transformed to

$$\frac{\partial \boldsymbol{u}^*}{\partial t} + \boldsymbol{u}^* \cdot \nabla \boldsymbol{u}^* = -\nabla p^* + \frac{1}{Re}\nabla \cdot (\nabla \boldsymbol{u}^* + \nabla \boldsymbol{u}^{*T}) + \frac{1}{We}\boldsymbol{F}^* + N(\boldsymbol{j}^* \times \boldsymbol{B}^*) + S^* \tag{2.14}$$

$$\nabla \cdot \boldsymbol{u}^* = 0 \tag{2.15}$$

It should be noted that without Marangoni effect, the surface tension force in the equation takes the non-dimensional form of

$$\frac{1}{We}\boldsymbol{F}^* = \frac{1}{We}\kappa^*\delta_s^*\boldsymbol{n}$$

however if Marangoni effect is included, it becomes

$$\frac{1}{We}\boldsymbol{F}^* = \frac{1}{Re}\left(\frac{1}{Ca}\kappa^*\delta_s^*\boldsymbol{n} + (T^* - T_0^*)\kappa^*\delta_s^*\boldsymbol{n} + \nabla_{\|}T^*\delta^*\right)$$

where the first two terms on the right-hand side (r.h.s) correspond to the isothermal normal stress balance and the temperature-dependent normal stress balance, and the third term corresponds to the Marangoni force.

The non-dimensional heat equation for the temperature is

$$\frac{\partial T^*}{\partial t} + \boldsymbol{u}^* \cdot \nabla T^* = \nabla \cdot \left(\frac{1}{Ma}\nabla T^*\right) \tag{2.16}$$

Therefore, the MHD flows exposed to an external MF are actually controlled by a series of dimensionless parameters. Without Marangoni effects, they correspond to

$$Re = \frac{\rho u_0 L_0}{\mu}, \; We = \frac{\rho u_0^2 L_0}{\sigma}, \; N = \frac{\sigma_e L_0 B_0^2}{\rho u_0}$$

while with the Marangoni effect, they are

$$Ca = \frac{\sigma_T G_T L_0}{\sigma_0}, \; Re = \frac{\rho \sigma_T G_T L_0^2}{\mu^2}, \; We = ReCa, \; Ma = \frac{\rho c_p \sigma_T G_T L_0^2}{\mu \lambda}$$

Re is the Reynolds number to represent the ratio between the inertial force and the viscous force, We the Weber number describing the ratio between the inertial force and the surface tension force, and N the interaction number being the ratio between the Lorentz force and the inertial force. Ca is the capillary number which gives the relative importance of the tangential to normal stresses at the interface. Note that the Marangoni number is equivalent to the Péclet number, for the characteristic velocity that is used in thermocapillary migration of drops or bubbles. In addition, another important non-dimensional parameter in describing the MHD flow is the Hartmann number, given as $Ha = \sqrt{ReN} = B_0 L_0 \sqrt{\frac{\sigma_e}{\mu}}$ to indicate the ratio between the Lorentz force and the viscous force. Particularly in the bubbly flows, whereas the gravitational force is dominant, the Froud number is used to describe the ratio between the inertial force and the buoyant force, being $Fr = \frac{u_0}{\sqrt{gL_0}}$.

2.3 Internal Boundary Conditions

In multi-phase MHD flows, the physical properties of ρ, μ, λ, c_p, σ_e are assumed to be constant within each phase, while they are discontinuous across the interface. Therefore, appropriate jump conditions should be satisfied across the interface for mass, momentum and energy balance considerations.

Some physical quantities should be continuous across the interface such as the velocity since no phase change occurs, which is represented as

$$[\![\mathbf{u}]\!] = 0 \tag{2.17}$$

where $[\![\,]\!]$ denotes the jump across the interface.

Similarly, The electric potential field and the temperature field are also continuous across the interface, yielding

$$[\![\varphi]\!] = 0, \quad [\![T]\!] = 0 \tag{2.18}$$

In the following part, $\boldsymbol{t}$ and $\boldsymbol{n}$ represent the tangent vector and normal vector to the frees surface, respectively.

The energy balance yields the continuity of the heat flux at the interface, while the conservation of charge leads the continuous normal component of the current density, both written as

$$[\![\lambda \nabla T]\!] \cdot \boldsymbol{n} = 0. \tag{2.19}$$

$$[\![\boldsymbol{J}]\!] \cdot \boldsymbol{n} = 0. \tag{2.20}$$

The stress along tangential direction at the interface between two fluids should be balanced as

$$\boldsymbol{t} \cdot [\![\mathbb{S}]\!] \cdot \boldsymbol{n} + \boldsymbol{t} \cdot [\![\mathbb{T}_l]\!] \cdot \boldsymbol{n} = \boldsymbol{F}_s \cdot \boldsymbol{t} \tag{2.21}$$

and in the normal direction as

$$[\![p]\!] + \boldsymbol{n} \cdot [\![\mathbb{S}]\!] \cdot \boldsymbol{n} + \boldsymbol{n} \cdot [\![\mathbb{T}_l]\!] \cdot \boldsymbol{n} = \boldsymbol{F}_s \cdot \boldsymbol{n} \tag{2.22}$$

As a consequence, the pressure field is not continuous in vicinity of the interface because of the surface tension and the normal electromagnetic stresses.

References

1. Müller U, Bühler L (2013) Magnetofluiddynamics in channels and containers. Springer Science and Business Media, Berlin
2. Ni MJ, Munipalli R, Morley NB et al (2007) A current density conservative scheme for incompressible MHD flows at a low magnetic Reynolds number. Part I: on a rectangular collocated grid system. J Comput Phys 227(1):174–204

Chapter 3
Numerical Schemes

3.1 The Discretisation of the Single-Phase MHD Equations

In Gerris, an approximate projection method making use of a staggered-in-time discretization is employed to discretize the governing equations, and the numerical schemes are second-order accurate. Combining the complete Navier–Stokes equations of Eq. 2.1 with the electromagnetic field equation of Eq. 2.11, a discretized form from time level k to $k+1$ is presented for the single-phase MHD flows:

$$\boldsymbol{J}^k = \sigma_e^{k+\frac{1}{2}}(-\nabla\varphi^k + \boldsymbol{u}^k \times \boldsymbol{B}^k) \tag{3.1}$$

$$\rho^{k+\frac{1}{2}}\left(\frac{\boldsymbol{u}^* - \boldsymbol{u}^k}{\Delta t} + \boldsymbol{u}^{k+\frac{1}{2}} \cdot \nabla\boldsymbol{u}^{k+\frac{1}{2}}\right) = -\nabla p^k + \nabla \cdot (\mu^{k+\frac{1}{2}}(\mathbb{D}^k + \mathbb{D}^*)) + (\boldsymbol{J} \times \boldsymbol{B})^k \tag{3.2}$$

$$\boldsymbol{u}^{k+1} = \boldsymbol{u}^* - \frac{\Delta t}{\rho^{k+\frac{1}{2}}}\nabla(p^{k+1} - p^k) \tag{3.3}$$

$$\nabla \cdot \boldsymbol{u}^{k+1} = 0 \tag{3.4}$$

$\boldsymbol{u}^*$ is the intermediate velocity calculated from the prediction step of Eq. 3.2, however, it does not meet the convergence-free condition. Therefore, a correction step is required to project u^* onto a divergence-free vector space by Eq. 3.3, and a Poisson equation is obtained to calculate the pressure difference in succession

$$\nabla \cdot \left(\frac{\Delta t}{\rho^{k+\frac{1}{2}}}\nabla(p^{k+1} - p^k)\right) = \nabla \cdot \boldsymbol{u}^* \tag{3.5}$$

J. Zhang, *The Developments and the Applications of the Numerical Algorithms in Simulating the Incompressible Magnetohydrodynamics with Complex Boundaries and Free Surfaces*, Springer Theses, https://doi.org/10.1007/978-981-10-6340-4_3

In the prediction step of Eq. 3.2, the advection term $\boldsymbol{u}^{k+\frac{1}{2}} \cdot \nabla \boldsymbol{u}^{k+\frac{1}{2}}$ is evaluated by using a conservative formulation developed from the Bell-Colella-Glaz second order unsplit upwind scheme [1]. On the other hand, a quad/octree based grid system is adopted for spatial discretization [7, 18], which is proved to be simple to implement a fully flexible adaptive mesh refinement strategy. The mesh can be adapted at every time step according to a mesh coarsening or refining criterion. The primary variables of the MHD equations, such as the velocity $\boldsymbol{u}$, the pressure p and the electric potential φ are located in the center of each finite volume. More detailed description about the numerical schemes can be found in the Ref. [18], and they will not be repeated here.

According to the governing equations described in Sect. 2.1, the solution of the electromagnetic field seems to be simple as it appears, however, hidden behind is an extra Poisson equation for the electric potential at each computation cycle. In order to accelerate the convergence rate, the multigrid technique is adopted that it is highly efficient when using quad/octrees grid system.

According to the electric charge conservation law, we must enhance

$$\nabla \cdot \boldsymbol{J}^k = 0 \tag{3.6}$$

Combined with Eq. 3.1, an electric potential Poisson equation [4] is derived at each time step as

$$\nabla \cdot (\sigma_e^{k+\frac{1}{2}} \nabla \varphi^k) = \nabla \cdot (\sigma_e^{k+\frac{1}{2}} (\boldsymbol{u}^k \times \boldsymbol{B}^k)) \tag{3.7}$$

As presented earlier, each discretized volume in domain is called a cell and all the variables are located in the cell center while the fluxes of the velocity and current density are located on the cell faces. Then a consistent and conservative scheme should be designed to ensure required accuracy as it is a true challenge [12, 22] in the last two decades. More recently, Ni et al. [16] proposed a consistent and conservative numerical scheme on the collocated grid system, and the numerical results were proved to be accurate and robust even if the Ha is as large as 10^4, which was frequently encountered in the tokamak device.

In the present study, we develop the consistent and conservative scheme to be adapted to the AMR framework with complex solid boundaries, irrespectively of electrically insulating or conducting walls.

3.1.1 *Cut-Cell Approach for MHD Flows with Electrically Insulating Boundaries*

For simpleness, the MHD flow without solid boundaries is viewed as a particular case of flows with complex boundaries, and only the solution procedure of the electromagnetic field is described as well, because the hydrodynamic part can be found in Ref. [18].

To begin with, the solid boundary is reconstructed by a serial of piecewise linear segments (2D) and the VOF method is used to calculate the fluid volume fraction in the mixed-cells, which are cut by the solid boundary. To calculate the electromagnetic field, the estimation of the flux of the electric potential and the current density across cell faces are very important that a consistent scheme should be applied. Besides, the Hartmann layer and side layer in vicinity of the solid boundaries are relative thin with a thickness proportional to $1/Ha$ and $\sqrt{1/Ha}$, respectively. Therefore, particular treatment is required in the region close to the boundaries due to the difficulties in imposing accurate boundary conditions on the embedded solids within the Cartesian grids.

3.1.1.1 Consistent Interpolation Operator and Gradient Operator at Coarse/Fine Cell Faces

In fact, if we apply a divergence theorem on Eq. 3.7 in a discrete cell, the Poisson equation yields

$$\oint \sigma_e \nabla\varphi \cdot \boldsymbol{n} dS = \oint \sigma_e (\boldsymbol{u} \times \boldsymbol{B}) \cdot \boldsymbol{n} dS \tag{3.8}$$

Then the discretization form of Eq. 3.8 can be written as,

$$\sum_{f=1}^{nf} (\sigma_e)_f \left(\frac{\partial\varphi}{\partial n}\right)_f S_f = \sum_{f=1}^{nf} (\sigma_e)_f (\boldsymbol{u} \times \boldsymbol{B})_f \cdot \boldsymbol{n}_f S_f \tag{3.9}$$

where f denotes the cell face (interface as well if the cell is cut by solid boundaries), $\boldsymbol{n}$ is the unit normal of the cell face, and S_f is the area of the face. nf, denoting the number of surrounding faces, is 4 in the 2D case and 6 in the 3D case. By assuming the length of the cell edge to be h and the fluid fraction to be s_f, we obtain $S_f = s_f h^2$ and a simplified equation is gained as below

$$\sum_{f=1}^{nf} (\sigma_e)_f s_f \left(\frac{\partial\varphi}{\partial n}\right)_f = \sum_{f=1}^{nf} (\sigma_e)_f s_f (\boldsymbol{u} \times \boldsymbol{B})_f \cdot \boldsymbol{n}_f \tag{3.10}$$

Usually, in order to keep the calculating of the electric potential flux across the partial cell face to be second-order accurate, a simple interpolating scheme from the surrounding cell-centers should be applied [5, 11, 23]. However, it is more difficult within the AMR framework that the consistent property is not easily retained, as a consequence, the full-face-center gradient and interpolation scheme is used here, and the solution decrease to first-order accuracy near the solid boundaries.

On uniform grids, it is usually not difficult to design a consistent scheme for the calculation of the current density on the cell faces, however it is more complex

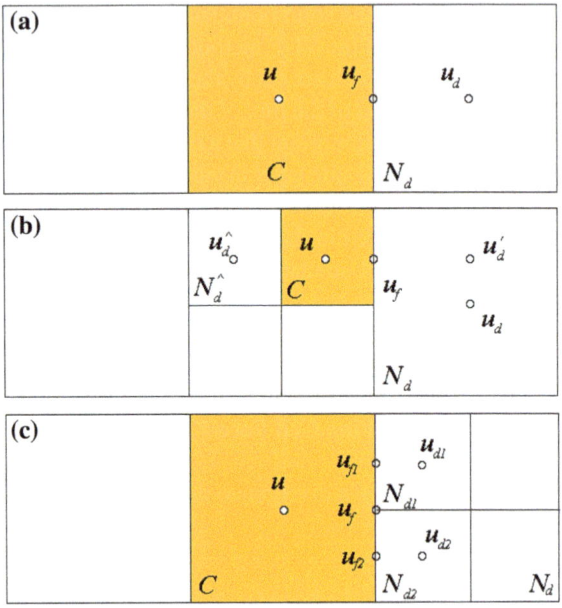

Fig. 3.1 Three cases for face-centered variable interpolation

within the AMR framework, because the mesh sizes may vary between neighbor cells, indicated by the coarse/fine cell face hereafter, and inconsistency may happen if improper interpolation scheme is adopted.

In practice, three cases must be considered in constructing the interpolation operator, and velocity interpolation is taken for example in Fig. 3.1.

Situation 1: as shown in Fig. 3.1a, if the neighbor cell N_d in direction d is at the same level as C and it is a leaf cell, the interpolated value from cell-centers is simply $u_f = (u + u_d)/2$.

Situation 2: as shown in Fig. 3.1b, if the neighbor cell N_d in direction d is at a lower level as C, a three point interpolating scheme is used to preserve the second-order accuracy on cell face, given by

$$u_f = u + \frac{1}{5}u'_d - \frac{1}{5}u_{\hat{d}} \tag{3.11}$$

where $u_{\hat{d}}$ is interpolated from u_d and its neighbouring cells in directions perpendicular to d.

Situation 3: as shown in Fig. 3.1c, if the neighbour cell N_d is at the same level of C but it is not a leaf cell, an ordinary interpolating scheme of $u_f = (u + u_d)/2$ is not consistent any more because the flux of u across the cell face calculated from cell C is $u_f S_f$ while it is $(u_{f1} S_{f1} + u_{f2} S_{f2})h/2$ by calculating from cell N_{d1} and N_{d2}, and obviously they are not equal. Therefore the estimation of u_f needs to take the average value of u_{f1} and u_{f2} to enhance the conservative and consistent property of flux across the cell face.

The above approach gives a consistent formula of the interpolation operator, however, it is not sufficient yet to calculate the righthand side of Eq. 3.10. Moreover, in order to keep $J_{nu\,f} = (\sigma_e)_f(\boldsymbol{u} \times \boldsymbol{B})_f \cdot \boldsymbol{n}_f$ be consistent on coarse/fine cell boundaries, we should treat the flux of $J_{nu\,f}$ as a whole. As similar as being described in *Situation* 3, the flux of $J_{nu\,f}$ on coarse/fine face should be calculated as $J_{nu\,f} = (J_{nu\,f1}S_{f1} + J_{nu\,f2}S_{f2})/(S_{f1} + S_{f2})$.

We define an operator of $flux(C, f, J_{nu})$ as the flux of J_{nu} across the face f of Cell C, and we also define *interpolate1*$(f, \boldsymbol{u})$, *interpolate2*$(f, \boldsymbol{u})$ as the interpolation operator of $\boldsymbol{u}$ in *Situation* 1 and *Situation* 2 respectively. Then the consistent scheme can be summarised as,

switch(face-type)
case(fine-fine)
$flux(C, f, J_{nu}) = (\sigma_e)_f(interpolate1(f, \boldsymbol{u}) \cdot interpolate1(f, \boldsymbol{B})) \cdot \boldsymbol{n}_f$
break
case(fine-coarse)
$flux(C, f, J_{nu}) = (\sigma_e)_f(interpolate2(f, \boldsymbol{u}) \cdot interpolate2(f, \boldsymbol{B})) \cdot \boldsymbol{n}_f$
break
case(coarse-fine)
$flux(C, f, J_{nu}) = flux(N_{d1}, f_{d1}, J_{nu}) + flux(N_{d2}, f_{d2}, J_{nu})$
break

In addition, to discrete the left hand side of Eq. 3.10, the calculation of the flux of $(J_{n\varphi})_f = -(\sigma_e)_f(\nabla\varphi)_f \cdot \boldsymbol{n}_f$ on coarse/fine cell faces also requires special treatment and the numerical method is similar with constructing the interpolation operator. Although the discretisation form is already introduced in Gerris, we would still repeat it here:

Situation 1: as shown in Fig. 3.1a, if the neighbor cell N_d in direction d is at the same level as C and it is a leaf cell, the flux of the gradient is given as:

$$(J_{n\varphi})_f = -(\sigma_e)_f\left(\frac{\partial\varphi}{\partial n}\right)_f = -(\sigma_e)_f\frac{\varphi_d - \varphi}{h} \tag{3.12}$$

Situation 2: as shown in Fig. 3.1b, if the neighbor cell N_d in direction d is at a lower level as C, the flux is given as:

$$(J_{n\varphi})_f = -(\sigma_e)_f\left(\frac{-5\varphi - 3\varphi_{\hat{d}} + 8\varphi'_d}{15h}\right) \tag{3.13}$$

where φ'_d is interpolated from φ_d and its neighboring cell in directions perpendicular to d.

Situation 3: as shown in Fig. 3.1c, if the neighbor cell N_d is at the same level of C but it is not a leaf cell, the flux $(J_{n\varphi})_f$ on the coarse/fine cell face is estimated as

$$(J_{n\varphi})_f = \frac{(J_{n\varphi})_{f1} S_{f1} + (J_{n\varphi})_{f2} S_{f2}}{S_{f1} + S_{f2}} \tag{3.14}$$

with $(J_{n\varphi})_{f1}$ and $(J_{n\varphi})_{f2}$ being calculated from cell N_{d1} and N_{d2} based on Eq. 3.13, and it sheerly enhances the conservative and consistent property of the flux across the cell face.

When the cell is cut by the electrically insulating solid boundary, both of $(J_{n\varphi})_f$ and $(J_{nu})_f$ are zero.

By substituting the discretized operators into Eq. 3.10, an efficient multigrid method is used to solve the Poisson equation and the electric potential is then obtained in the cell-center. After that, according to Ohm's law, the current density is calculated on the cell faces. However, it should be kept in mind that the gradient operator used here need to be consistent with that used in solving the electric potential Poisson equation. Moreover, to better enhance the consistency in calculating the current density on cell faces, an average scheme is adopted again on the coarse/fine cell faces, given by

$$(J_n)_f = \frac{(J_n)_{f1} S_{f1} + (J_n)_{f2} S_{f2}}{S_{f1} + S_{f2}} \tag{3.15}$$

3.1.1.2 Calculation of the Lorentz Force at the Cell Center

Since the current density fluxes j_x, j_y, j_z are located on the cell faces in normal directions, and therefore, interpolations from the cell faces to the cell center are needed to calculate the Lorentz force.

For a mixed-cell, a simple interpolation operator for J based on the area average scheme is given as,

$$(j_i)_c = \frac{\sum_{f=1}^{nf} (j_i |s_i|)_f}{\sum_{f=1}^{nf} |s_i|_f}$$

where i denotes the x-, y-, z-directions, and $|s_i|_f$ is the area of the partial face covered by the fluid. This scheme is adopted in Gerris [18] for the interpolation of the pressure gradient. However, as depicted by Ni et al. [17], this scheme introduces numerical error for the calculation of the Lorentz force on an irregular mesh. Therefore, in the mixed-cells, a conservative formulation developed by Ni et al. [17] is adopted.

$$\begin{aligned}
\boldsymbol{J}_c &= \frac{1}{\Omega_c}\int_{\Omega_c} \boldsymbol{J} d\Omega = \frac{1}{\Omega_c}\int_{\Omega_c} \nabla \cdot (\boldsymbol{J}(\boldsymbol{r} - \boldsymbol{r}_c))d\Omega \\
&= \frac{1}{\Omega_c}\oint_{s_f} J_n(\boldsymbol{r} - \boldsymbol{r}_c)ds \\
&= \frac{1}{\Omega_c}\sum_{f=1}^{nf}(J_n)_f(\boldsymbol{r}_f - \boldsymbol{r}_c)s_f \\
&= \frac{1}{\Omega_c}\sum_{f=1}^{nf}(J_n)_f s_f d_f \boldsymbol{n}_f
\end{aligned} \tag{3.16}$$

Herein, Ω_c denotes the partial volume of the cell covered by fluid and s_f is the corresponding face fraction, and the weighting is given by $\boldsymbol{r}_f - \boldsymbol{r}_c = d_f \boldsymbol{n}_f$ where d_f denotes the distance from the centre of the partial cell to the centre of the partial face. If the cell is a regular square, then Eq. 3.16 can be reduced as,

$$\boldsymbol{J}_c = \sum_{f=1}^{nf}(J_n)_f \frac{s_f d_f \boldsymbol{n}_f}{\Omega_c} = \frac{1}{2}\sum_{f=1}^{nf}(J_n)_f \boldsymbol{n}_f \tag{3.17}$$

This is the ordinary averaged interpolation scheme from cell face to cell centre and it indicates that the simple interpolation scheme for current density calculation in a full square cell is conservative and consistent.

With the conservative interpolation operator of Eq. 3.16, the Lorentz force is calculated by $\boldsymbol{F}_l = \boldsymbol{J}_c \times \boldsymbol{B}_c$ in the mixed-cells.

3.1.2 Cut-Cell Approach for MHD Flow with Electrically Conducting Boundaries

As described in Sect. 1.2.1 when cut-cell method is applied to treat the ordinary flows within complex boundaries, the meshes inside the solids are always destroyed and abandoned. However, regarding the MHD flows with electrically conducting boundaries, the walls will seriously affect the distribution of the current density [9], and the flow characteristics are completely different from the flows bounded by the electrically insulating wall. Therefore, the computation of the electromagnetic field should also contain the solid walls in an electrically conducting case.

Grigoriadis et al. [8] introduced the immersed boundary method to simulate MHD flows with electrically insulating wall by applying a forcing term during calculating the current density to enhance $\boldsymbol{J} = 0$ on the fluid-solid boundaries. But this method can not be extended to the MHD flows bounded by electrically conducting walls directly, because there is no definite boundary condition along the rigid wall that the electric potential and current density are continuous across the solid boundary.

In this paper, we develop a conservative approach by using the VOF method to deal with the cut cells lying astride the fluid region and solid region, so as to make cut-cell method be applicable in MHD flows bounded by the electrically conducting walls. And this technique is based on the consistent and conservative scheme covered earlier in Sect. 3.1.1 for dealing with the electrically insulating walls.

3.1.2.1 The Calculation of the Electric Potential Poisson Equation

The difficulty in applying the cut-cell method, evidently, is to solve the variable coefficient Poisson equation with jump conditions in the cut-cell. Compared to the solution of the pressure Poisson equation, the coupling fluid-solid domain should be conjugated in solving the electric potential Poisson equation, and therefore, the destroyed cells in the solid should be reconstructed first and all the corresponding variables, such as p, $\boldsymbol{u}$, are copied to the newly generated grid system.

For instance, this problem is described in Fig. 3.2 that the mixed-cell (i, j) contains fluid and solid, the discretized integral form of the electric potential Poisson equations, given in Eq. 3.8, is formulated as:

$$\sum_{f=1}^{nf}((\sigma_e)_{f1}\left(\frac{\partial\varphi}{\partial n}\right)_{f1} S_{f1} + (\sigma_e)_{f2}\left(\frac{\partial\varphi}{\partial n}\right)_{f2} S_{f2}) = \sum_{f=1}^{nf}(\sigma_e)_{f1}(\boldsymbol{u}\times\boldsymbol{B})_f \cdot \boldsymbol{n}_f S_{f1} \quad (3.18)$$

where the superscripts 1 and 2 indicate the fluid and solid respectively, $S_1 = |CF|$, $S_2 = |FB|$ on face $|CB|$ in Fig. 3.2. When calculating $(\frac{\partial\varphi}{\partial n})_{f1}$ and $(\frac{\partial\varphi}{\partial n})_{f2}$ on the partial face, Johansen and Colella [11] suggested to evaluate the fluxes on the partial cell face based on a linear interpolation between adjacent fluxes, so $(\frac{\partial\varphi}{\partial n})_{f1}$ is obtained by using a quadratic polynomial based on the surrounding nodal values in the fluid part, and alternatively, $(\frac{\partial\varphi}{\partial n})_{f2}$ is evaluated from the surrounding solid nodes. Another

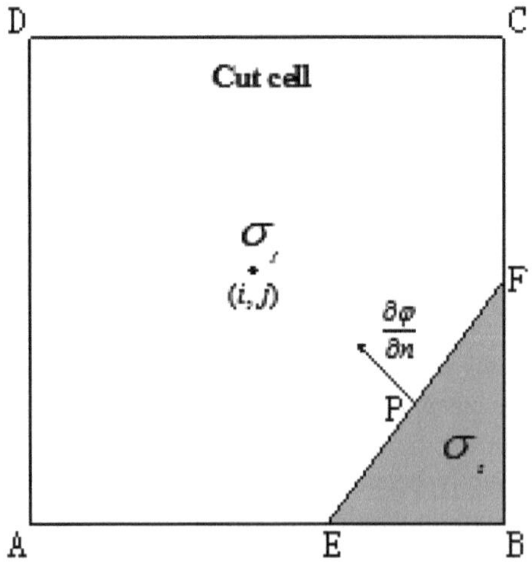

Fig. 3.2 The mixed-cell lying between the fluid and the electrically conducting wall

second-order accurate method for variable coefficient Poisson equation is presented by H. Li et al. [10], they calculate the solid part and the fluid part are calculated separately. Moreover, X.D. Liu et al. [14] introduced the Ghost Fluid Method (GFM) to capture the boundary conditions with a contact discontinuity, nevertheless, the jump condition across the interface in the present problem is not so definite. Referring to immersed boundary method, R.J. LeVeque et al. [13] developed a second-order numerical method to preserve the discontinuity across the internal boundary too.

However,as mentioned before, it is so complex to develop these interpolation schemes with the adaptive framework and the existing cell-centred pressure solvers could not be used directly to solve the poisson equation. Consequently, the gradient of the electric potential on the mixed-cell face is considered to be uniform, and the following formula is derived:

$$\sum_{f=1}^{nf}((\sigma_e)_{f1}S_{f1} + (\sigma_e)_{f2}S_{f2})\left(\frac{\partial \varphi}{\partial n}\right)_f = \sum_{f=1}^{nf}\sigma_1(\boldsymbol{u} \times \boldsymbol{B})_f \cdot \boldsymbol{n}_f S_{f1} \tag{3.19}$$

Moreover, if the solid is electrically insulating that $\sigma_2 = 0$, then Eq. 3.19 is simplified as,

$$\sum_{f=1}^{nf}(\sigma_e)_{f1}S_{f1}\left(\frac{\partial \varphi}{\partial n}\right)_f = \sum_{f=1}^{nf}\sigma_1(\boldsymbol{u} \times \boldsymbol{B})_f \cdot \boldsymbol{n}_f S_{f1} \tag{3.20}$$

which is consistent with Eq. 3.10 when the wall is electrically insulating.

3.1.2.2 Calculation of the Lorentz Force at the Cell Center

Even if the solid walls are electrically conducting, however, we just need to calculate the Lorentz force in the fluid part. Then a consistent and conservative interpolation scheme, as similar as Eq. 3.16, is adopted to calculate the current density from the cell faces to the cell center.

However, it should be noted that the boundary condition of the electric potential does not conform to $\frac{\partial \varphi}{\partial n} \neq 0$ on the solid walls because it is electrically conducting, and this is shown in Fig. 3.2. It should be noticed that the gradient of the electric potential through $|EF|$ calculated from the fluid part and the solid part are different due to the jump condition across the boundary, however, the flux should be the same. A simple correlation on the solid boundary is given by,

$$(\sigma_e)_f\left(\frac{\partial \varphi}{\partial n}\right)_f = (\sigma_e)_{f1}\left(\frac{\partial \varphi}{\partial n}\right)_{f1} = (\sigma_e)_{f2}\left(\frac{\partial \varphi}{\partial n}\right)_{f2} \tag{3.21}$$

Because the electric potential is continuous across the boundary, therefore, the calculation of $\frac{\partial \varphi}{\partial n}$ on the solid boundary uses a bilinear interpolation scheme from the surrounding nodes. The bilinear function in 2D case takes the form of

$$\varphi = ax + by + cxy + d \tag{3.22}$$

Then the coefficients are estimated by substituting the values of φ from the neighbor cells into this equation. $\frac{\partial \varphi}{\partial x}$ and $\frac{\partial \varphi}{\partial y}$ on $|EF|$ are also obtained, according to

$$\begin{aligned} \varphi_x &= a + cy_p \\ \varphi_x &= b + cx_p \end{aligned} \tag{3.23}$$

Furthermore, when evaluating the conductivity at the embedded boundary, the *weighted harmonic mean interpolation* scheme is proved to be accurate when treating such coupling problems,written as

$$\frac{1}{(\sigma_e)_f} = \frac{c}{(\sigma_e)_1} + \frac{1-c}{(\sigma_e)_2} \tag{3.24}$$

where c denotes the area ratio of the fluid in the mixed-cell and $1-c$ is the solid part accordingly. After that, the current density on interface is calculated as

$$(J_n)_f = (J_{n\varphi})_f = -(\sigma_e)_f \left(\frac{\partial \varphi}{\partial n}\right)_f \tag{3.25}$$

where $(J_{nu})_f$ turns to be zero on the boundary because of the no-slip condition.

Moreover, if we presume $\frac{\partial \varphi}{\partial n} = 0$ on the solid boundary, the method is consistent with what we presented in Sect. 3.1.1 for MHD flows bounded by electrically insulating boundaries.

3.2 The Discretisation of the Multi-phase MHD Equations

Regarding the original multi-phase flows without MHD effect or Marangoni effect, the numerical schemes are described in details by S. Popinet [19], and the fluid-fluid interface is advanced and reconstructed by the VOF method while the normal surface tension stress is treated as a continuous volume force using the CSF model as proposed by Brackbill et al. [2]. To reduce the spurious flows in the numerical simulations, a balanced-force technique proposed by Francois et al. [6] is also adopted to balance the pressure gradient and the surface tension force. Furthermore, an accurate height-function method is used to estimate the curvature since it is also important for surface tension calculation [3].

3.2.1 Without the Marangoni Effect

Within the VOF framework, a function of the volume fraction f is introduced to track the interface and if phase change is excluded, the conservation of a certain phase mass is represented as

$$\frac{\partial f}{\partial t} + \nabla \cdot (\boldsymbol{u} f) = 0 \tag{3.26}$$

with the calculated volume fraction, the physical properties of the fluids are updated as

$$\phi = f\phi_1 + (1 - f)\phi_2 \tag{3.27}$$

where ϕ represents any of ρ, μ, σ_e, λ and c_p of both fluids, the subscript 1 and 2 denote the disperse and the continuous phase, respectively. It should be noticed that in certain situations, a filtered scheme is used to smear the discontinuity of the physical properties, in order to obtain more stable numerical results.

When Marangoni effect is taken into consideration, as Eq. 7.1 shows, the surface tension is divided into two parts that the first part $\boldsymbol{F}_s^n$ is normal to the interface while the second part $\boldsymbol{F}_s^t$ is along the tangent direction. And with the proposed CSF model [2], the normal component in Eq. 7.1 takes the form of

$$\sigma\kappa\delta_s\boldsymbol{n} = \sigma\kappa \parallel \nabla f \parallel \frac{\nabla f}{\parallel \nabla f \parallel} = \sigma\kappa\nabla f \tag{3.28}$$

Besides, Eq. 2.22 indicates the normal part should be balanced by the pressure gradient at interface which is the central idea of the balanced-force technique [6], and hence it is calculated at the same location as the pressure gradient. In addition, Eq. 2.21 describes if the Lorentz force is eliminated, any mechanism which maintains $\boldsymbol{F}_s^t \neq 0$ will drive the static fluid moving, and therefore, the tangential part of the surface tension force is discretized at the same location with the viscous stress.

Compared with the numerical schemes designed for the single phase MHD flows, i.e. from Eqs. 3.1 to 3.5, there are some additional numerical steps regarding to the multi-phase MHD flows. First, a VOF method is employed to calculate the volume fraction $f_c^{k+\frac{1}{2}}$ of different fluids at the intermediate time step, and it is simply expressed as:

$$\frac{f_c^{k+\frac{1}{2}} - f_c^{k-\frac{1}{2}}}{\Delta t} + \nabla_c \cdot (f^k \boldsymbol{u}^k) = 0 \tag{3.29}$$

It should be noted that the advection of f in Eq. 3.29 is much more complicated than it appears, an directional-split Eulerian - implicit C explicit scheme described by Rider and Kothe [20] is adopted in Gerris, and the volume fluxes are estimated using the geometry of the reconstructed interface based on the PLIC scheme [21, 24], which is efficient and simple to implement on Cartesian grids. However, there are some difficulties in adapting this scheme to more general spatial discretisations when the interface moves through a coarse/fine cell face, and a more detailed description is introduced by S. Popinet [19].

Afterwards, all physical properties are updated by

$$\phi_c^{k+\frac{1}{2}} = f_c^{k+\frac{1}{2}}\phi_1 + (1 - f_c^{k+\frac{1}{2}})\phi_2 \tag{3.30}$$

Regarding the temperature field, it is also calculated at the intermediate time step using $\boldsymbol{u}^{k+\frac{1}{2}}$ which is advected by the Bell–Colella–Glaz scheme [1].

$$\frac{T_c^{k+\frac{1}{2}} - T_c^{k-\frac{1}{2}}}{\Delta t} + (\boldsymbol{u}\cdot\nabla T)_c^{k+\frac{1}{2}} = \nabla_c \cdot \left(\left(\frac{\lambda}{\rho c_p}\right)_f^{k+\frac{1}{2}} \nabla_f T_c^{k+\frac{1}{2}}\right) \tag{3.31}$$

Afterwards, the temperature is used to evaluate the Marangoni force, which will be discussed in the following study. In the equations above, the subscript c denotes the cell center and f denotes the cell face. $\nabla_c\cdot$ means the divergence will be taken at a cell center based on the fluxes estimated on the cell faces, while ∇_f means the gradient will be discretized on a cell face. By using the balanced-force technique in coupling the pressure gradient and the surface tension force, the prediction step in solving the Navier–Stokes equations, namely Eq. 3.2, are transferred to

$$\begin{aligned}\frac{\boldsymbol{u}_c^* - \boldsymbol{u}_c^k}{\Delta t} + (\boldsymbol{u}\cdot\nabla\boldsymbol{u})_c^{k+\frac{1}{2}} =& \frac{1}{\rho_c^{k+\frac{1}{2}}}\nabla_c\cdot\left(\mu_f^{k+\frac{1}{2}}(\mathbb{D}^k + \mathbb{D}^*)_f\right) - \left(\frac{1}{\rho}\nabla p - \frac{1}{\rho}\sigma\kappa\nabla f\right)_c^{k-\frac{1}{2}} \\ &+ \left(\frac{1}{\rho}\sigma_T\nabla_{\parallel}T\delta_s\right)_c^{k+\frac{1}{2}} + \frac{1}{\rho_c^{k+\frac{1}{2}}}(\boldsymbol{J}\times\boldsymbol{B})_c^{k+\frac{1}{2}} + \boldsymbol{g}\end{aligned} \tag{3.32}$$

After that, the provisional value of $\boldsymbol{u}_c^*$ is used to evaluate the velocity at next time step $k+1$ in cell centers. Accordingly, the correction step of the approximate projection method is formulated as

$$\boldsymbol{u}_c^{k+1} = \boldsymbol{u}_c^* - \Delta t\left(\left(\frac{1}{\rho}\nabla p - \frac{1}{\rho}\sigma\kappa\nabla f\right)_c^{k+\frac{1}{2}} - \left(\frac{1}{\rho}\nabla p - \frac{1}{\rho}\sigma\kappa\nabla f\right)_c^{k-\frac{1}{2}}\right) \tag{3.33}$$

On a collocated mesh, the pressure gradient on a cell face is needed to overcome the checkerboard instability for the incompressible flows. To ensure the balance between the normal surface tension and the pressure gradient, a consistent technique is usually needed to reduce the spurious flow that Francois et al. [6] and Ni [15] indicate the computational accuracy would be improved if the pressure gradient at the cell center is interpolated from the cell faces instead of directly using a second-order central difference scheme from the neighbor cells surrounding it:

$$\left(\frac{1}{\rho}\nabla p - \frac{1}{\rho}\sigma\kappa\nabla f\right)_c = \left(\frac{1}{\rho}\nabla p - \frac{1}{\rho}\sigma\kappa\nabla f\right)_{f\mapsto c} \tag{3.34}$$

After being estimated on the cell faces, the normal surface tension is added to the face-located provisional velocity, which is interpolated as $\boldsymbol{u}_f^* = \boldsymbol{u}_{c\mapsto f}^*$. After that, the pressure Poisson equation is computed though the multilevel solver.

$$\nabla_c \cdot \left(\frac{1}{\rho_f^{k+\frac{1}{2}}}\nabla_f p^{k+\frac{1}{2}} - \frac{1}{\rho_f^{k-\frac{1}{2}}}\nabla_f p^{k-\frac{1}{2}}\right) = \frac{1}{\Delta t}\nabla_c \cdot \boldsymbol{u}_f^{**} \tag{3.35}$$

where $\boldsymbol{u}_f^{**}$ indicates the provisional velocity updated by adding the surface-tension force before projection steps. That is, $\boldsymbol{u}^{**}$ located on the cell faces has the following formulation

$$\boldsymbol{u}_f^{**} = \boldsymbol{u}_f^* + \Delta t\left(\frac{1}{\rho_f^{k+\frac{1}{2}}}(\sigma\kappa\nabla f)_f^{k+\frac{1}{2}} - \frac{1}{\rho_f^{k-\frac{1}{2}}}(\sigma\kappa\nabla f)_f^{k-\frac{1}{2}}\right) \tag{3.36}$$

The above algorithm was proposed by Francois et al. [6] and they concluded this to be crucial to decrease the spurious flows. However, Ni [15] proved that the key issue was to use a consistent scheme to discrete the pressure gradient and the surface tension force, no matter whether they were calculated at the same location or not. Therefore, we put Marangoni force in Eq. 3.32 but not in Eq. 3.33 that it should be balanced by the viscous stress in the tangential direction.

At last, The cell-centered velocity $\boldsymbol{u}_c^{k+1}$ is updated by Eq. 3.33 after solving the pressure Poisson equation.

3.2.2 With the Marangoni Effect

As depicted in Eq. 3.32, the calculation of the Marangoni forces requires a priori unknown temperature gradient tangent to the interface, which is relevant to the curvature of the free surface. Furthermore, the singular distribution of the Marangoni force in vicinity the interface also complicates the numerical computations.

For the first difficulty, the reconstructed geometrical interface based on the AMR technique enables to solve it. For another respect regarding the singularity of the Marangoni force, the idea of the CSF model is introduced as being previously used in calculating the normal surface tension force.

The tangent gradient operator along the interface is calculated firstly, given by

$$\nabla_{\|} = (\boldsymbol{I} - \boldsymbol{n}\otimes\boldsymbol{n})\nabla \tag{3.37}$$

where $\boldsymbol{I}$ is the identity matrix and $(\boldsymbol{n}\otimes\boldsymbol{n})$ is the dyadic product of the normal vector with itself. The tensor $(\boldsymbol{I} - \boldsymbol{n}\otimes\boldsymbol{n})$ singles out the tangential projection of a vector.

Then

$$\nabla_{\parallel} T = (\boldsymbol{I} - \boldsymbol{nn})\nabla T = \nabla T - (\nabla T \cdot \boldsymbol{n})\boldsymbol{n} \tag{3.38}$$

$$\sigma_T \nabla_{\parallel} T \delta_s = \sigma_T (\nabla T - (\nabla T \cdot \boldsymbol{n})\boldsymbol{n})\delta_s \tag{3.39}$$

where the normal vector of the interface is evaluated as $\boldsymbol{n} = \frac{\nabla f}{|\nabla f|}$, and the delta function δ_s is also replace by the smeared form as $\delta_s =| \nabla f |$. Thus the Marangoni force is transformed to a bulk volume force, written as

$$\boldsymbol{F}_s^t = \sigma_T \left(\nabla T - \left(\nabla T \cdot \frac{\nabla f}{| \nabla f |} \right) \frac{\nabla f}{| \nabla f |} \right) | \nabla f | \tag{3.40}$$

As the Marangoni stresses need to be balanced by the viscous stresses at the tangent direction of the interface, Eq. 2.21 is expressed as below if Lorentz force is not included

$$\mu_2 \left(t_i \frac{\partial u_i}{\partial n} + n_k \frac{\partial u_k}{\partial s} \right)_2 - \mu_1 \left(t_i \frac{\partial u_i}{\partial n} + n_k \frac{\partial u_k}{\partial s} \right)_1 = \sigma_T \frac{\partial T}{\partial s} \tag{3.41}$$

where t_i, n_k is the tangential direction and the normal direction respectively. $\frac{\partial}{\partial s}$, $\frac{\partial}{\partial n}$ are the surface derivative and normal derivative respectively.

Motivated by this, the Marangoni force is calculated to be consistent with the discretisation of the viscous force. That is to say, the discretisation is firstly performed on cell faces and then interpolated to the cell center. To illustrate this in details, the interpolation scheme in x-direction is presented as an example. The simple interpolation on a non-rectangular cell is given as

$$(F_x)_c = \frac{\sum_{f=1}^{nf} (F_x \mid s_x \mid)_f}{\sum_{f=1}^{nf} \mid s_x \mid_f} \tag{3.42}$$

where s is the area of the cell face normal to x-direction, c and f represents cell center and face center respectively. When the cell is not cut by solid, Eq. 3.42 is transformed to

$$(F_x)_c = \frac{(F_x)_{f1} + (F_x)_{f2}}{2} \tag{3.43}$$

Before that, the evaluation of Marangoni force on the cell faces is obtained by

$$\begin{aligned} (F_x)_f &= \sigma_T \left(\nabla T - \left(\nabla T \cdot \frac{\nabla f}{| \nabla f |} \right) \frac{\nabla f}{| \nabla f |} \right)_f | \nabla_f f | \cdot \boldsymbol{i}_x \\ &= \sigma_T \left(\left(\frac{\partial T}{\partial x} \right)_f | \nabla_f f | - \left(\frac{\nabla_f T \cdot \nabla_f f}{| \nabla_f f |} \right) \left(\frac{\partial f}{\partial x} \right)_f \right) \end{aligned} \tag{3.44}$$

where $\left(\frac{\partial}{\partial x}\right)_f$ is calculated directly because x- is normal to the face. When the neighbor cells separated by f are at the same level, a simple central difference scheme is used to calculate $\left(\frac{\partial}{\partial x}\right)_f$, nevertheless, a three point interpolation scheme is required if the neighbor cells are at different levels.

When estimating other components of ∇_f at the cell face normal to x-direction, it is more complicated since $\frac{\partial}{\partial y}$ and $\frac{\partial}{\partial z}$ can not be calculated directly. We take the 2D case for an example while the extension to the 3D case is straightforward.

$$\left(\frac{\partial T}{\partial y}\right)_f = \frac{(T_{f,y+} - T_{f,y-})}{\Delta y} \tag{3.45}$$

where $T_{f,y+}$ and $T_{f,y-}$ are the values on the cell vertexes in 2D case and on the edge centers in 3D case, respectively. Obviously, $T_{f,y+}$ and $T_{f,y-}$ need to be interpolated from the surround cell centers in 2D case, while for the 3D case, the edge center value needs an additional interpolation from the cell vertexes.

By applying the interpolation operators, $\boldsymbol{F}_s^t$ is calculated on the cell faces first and then it is interpolated to the cell center. Moreover, within the adaptive framework that the neighbor cells may vary in size, the estimation of the Marangoni forces on the coarse/fine cell faces also needs to be treated in a consistent manner, which is described in details in Sect. 3.1.1.1.

3.3 Summary

A consistent and conservative scheme for calculating the electric potential and the current density is proposed to simulate the incompressible MHD flows, respectively with complex boundaries or free surfaces. The numerical algorithms are developed as an extension of the Gerris solver, which involves the AMR technique and the cut-cell method.

When solving the single-phase MHD flows, the thin boundary layers are exactly solved by using fine grids while coarse meshes are still used in the central region of the flow, and therefor the computational efficiency is expected to be improved. The cut-cell approach for solving the ordinary flows has been extended to simulate the MHD flows with complex boundaries. Moreover, a VOF method based on the Cartesian grids is combined to simulate the MHD flows bounded by the electrically conducting walls. When the meshes are no longer uniform in the adaptive framework, it is crucial to ensure the calculation of the current density flux to be conservative on the coarse/fine cell faces. In particular, the estimation of $(\sigma_e \frac{\partial \varphi}{\partial n})_f$ and $(\sigma_e(\boldsymbol{u} \times \boldsymbol{B}))_f \cdot \boldsymbol{n}_f$ on the coarse/fine cell faces when solving the electric potential Poisson equations are presented in a consistent formulation. Base on the Cartesian grids, numerically simulating MHD flows with electrically conducting walls is more difficult than that with electrically insulating walls, because electromagnetic parameters are not continuous across the solid-fluid boundaries in mixed-cells. VOF method is proved to be efficient in dealing with this problem in one grid system.

Regarding the multi-phase MHD flows, the volume of fluid method is employed to capture the interface accurately between the two fluids. The treatment of the surface tension force is as the same as those in the original Gerris code. Besides, an accurate method to estimate the Marangoni force is developed combined with the CSF surface tension scheme and a consistent technique is also implemented to calculate the flux across the coarse/fine cell faces.

References

1. Bell JB, Colella P, Glaz HM (1989) A second-order projection method for the incompressible Navier-Stokes equations. J Comput Phys 85(2):257–283
2. Brackbill JU, Kothe DB, Zemach C (1992) A continuum method for modeling surface tension. J Comput Phys 100(2):335–354
3. Cummins SJ, Francois MM, Kothe DB (2005) Estimating curvature from volume fractions. Comput Struct 83(6):425–434
4. Davidson PA (2001) An introduction to magnetohydrodynamics. Cambridge University Press, Cambridge
5. Fadlun EA, Verzicco R, Orlandi P et al (2000) Combined immersed-boundary finite-difference methods for three-dimensional complex flow simulations. J Comput Phys 161(1):35–60
6. Francois MM, Cummins SJ, Dendy ED et al (2006) A balanced-force algorithm for continuous and sharp interfacial surface tension models within a volume tracking framework. J Comput Phys 213(1):141–173
7. Fryxell B, Olson K, Ricker P et al (2000) FLASH: an adaptive mesh hydrodynamics code for modeling astrophysical thermonuclear flashes. Astrophys J Suppl Ser 131(1):273
8. Grigoriadis DGE, Kassinos SC, Votyakov EV (2009) Immersed boundary method for the MHD flows of liquid metals. J Comput Phys 228(3):903–920
9. Hunt JCR (1965) Magnetohydrodynamic flow in rectangular ducts. J Fluid Mech 21(4):577–590
10. Ji H, Lien FS, Yee E (2006) An efficient second-order accurate cut-cell method for solving the variable coefficient Poisson equation with jump conditions on irregular domains. Int J Numer Methods Fluids 52(7):723–748
11. Johansen H, Colella P (1998) A Cartesian grid embedded boundary method for Poisson's equation on irregular domains. J Comput Phys 147(1):60–85
12. Leboucher L (1999) Monotone scheme and boundary conditions for finite volume simulation of magnetohydrodynamic internal flows at high Hartmann number. J Comput Phys 150(1):181–198
13. Leveque RJ, Li Z (1994) The immersed interface method for elliptic equations with discontinuous coefficients and singular sources. SIAM J Numer Anal 31(4):1019–1044
14. Liu XD, Fedkiw RP, Kang M (2000) A boundary condition capturing method for Poisson's equation on irregular domains. J Comput Phys 160(1):151–178
15. Ni MJ (2009) Consistent projection methods for variable density incompressible Navier-Stokes equations with continuous surface forces on a rectangular collocated mesh. J Comput Phys 228(18):6938–6956
16. Ni MJ, Munipalli R, Morley NB et al (2007) A current density conservative scheme for incompressible MHD flows at a low magnetic Reynolds number. Part I: on a rectangular collocated grid system. J Comput Phys 227(1):174–204
17. Ni MJ, Munipalli R, Huang P et al (2007) A current density conservative scheme for incompressible MHD flows at a low magnetic Reynolds number. Part II: on an arbitrary collocated mesh. J Comput Phys 227(1):205–228

18. Popinet S (2003) Gerris: a tree-based adaptive solver for the incompressible Euler equations in complex geometries. J Comput Phys 190(2):572–600
19. Popinet S (2009) An accurate adaptive solver for surface-tension-driven interfacial flows. J Comput Phys 228(16):5838–5866
20. Rider WJ, Kothe DB (1998) Reconstructing volume tracking. J Comput Phys 141(2):112–152
21. Scardovelli R, Zaleski S (1999) Direct numerical simulation of free-surface and interfacial flow. Annu Rev Fluid Mech 31(1):567–603
22. Sterl A (1990) Numerical simulation of liquid-metal MHD flows in rectangular ducts. J Fluid Mech 216:161–191
23. Ye T, Mittal R, Udaykumar HS et al (1999) An accurate Cartesian grid method for viscous incompressible flows with complex immersed boundaries. J Comput Phys 156(2):209–240
24. Youngs DL (1982) Time-dependent multi-material flow with large fluid distortion. Academic Press, New York, Numerical Methods for Fluid Dynamics

Chapter 4
The Validations of the Numerical Methodology

In this chapter, detailed validations are presented to investigate the accuracy and the efficiency of the Gerris-MHD solver in simulating the MHD flows, respectively with complex solid boundaries or free surfaces.

4.1 The Single-Phase MHD Flows with Complex Boundaries

4.1.1 Convergence for a Simple Periodic Problem

Firstly, we would like to test the consistent property of the numerical scheme in calculating $\boldsymbol{J}$ on the coarse/fine cell faces in solving the electric potential Poisson equation, which is of great importance in preserving the charge conservative law. Besides, the numerical results based on an inconsistent scheme are also presented for comparison.

For the ordinary flows, there are different numerical cases to test the accuracy of the algorithms. However, for MHD flows, there is no available unsteady problem with exact analytic solution being proposed and extensive used in available literatures. Herein, we consider a simple periodic problem in a square unit domain which has been widely studied in ordinary flows [1, 10, 16]. The initial conditions are set as

$$u(x, y) = 1 - 2cos(2\pi x)sin(2\pi y),$$
$$v(x, y) = 1 + 2sin(2\pi x)cos(2\pi y). \tag{4.1}$$

where the boundary conditions in x- and y-directions are both periodic, then the exact solutions of the Euler equations are derived as

J. Zhang, *The Developments and the Applications of the Numerical Algorithms in Simulating the Incompressible Magnetohydrodynamics with Complex Boundaries and Free Surfaces*, Springer Theses, https://doi.org/10.1007/978-981-10-6340-4_4

$$
\begin{aligned}
u(x, y, t) &= 1 - 2cos(2\pi(x - t))sin(2\pi(y - t)),\\
v(x, y, t) &= 1 + 2sin(2\pi(x - t))cos(2\pi(y - t)),\\
p(x, y, t) &= -cos(4\pi(x - t)) - cos(4\pi(y - t)).
\end{aligned} \tag{4.2}
$$

In order to make it be applicable for the MHD flows, a uniform MF is imposed perpendicular to the flow field, given as B_z, and the Lorentz force is always zero in such a problem. Then the electric potential is estimated as

$$
\begin{aligned}
\nabla^2\varphi &= \nabla \cdot (\boldsymbol{u} \times \boldsymbol{B})\\
&= \frac{\partial(vB_z)}{\partial x} - \frac{\partial(uB_z)}{\partial y}\\
\Rightarrow \varphi &= -\frac{1}{\pi}cos(2\pi(x - t))cos(2\pi(y - t)) + c
\end{aligned} \tag{4.3}
$$

where c is an arbitrary constant and is taken as the average value of the computed electric potential in the flow domain.

Since our purpose is to investigate whether the calculation of the electric potential is consistent on the coarse/fine cell faces, the consistent interpolation scheme proposed in Sect. 3.1.1.1 is named as *Scheme I*. By contrary, if the flux on coarse/fine cell faces are calculated separately from the neighbor cells, that means *Situation 3* in Sect. 3.1.1.1 is not considered particularly, it is viewed as an inconsistent scheme of *Scheme II*.

This test is firstly conducted on the uniform grids with $L = 5$, 6 and 7 levels of refinement and then one or two additional levels are added in a square domain defined by the points $(-0.25, -0.25)$ and $(0.25, 0.25)$. All simulations are performed at the same CFL number of 0.75 and the running time for each is 0.5.

When *Scheme I* is applied in calculating $\boldsymbol{J}_{nf}$ on cell faces, Fig. 4.1 shows the spatial evolution of the L_2 and L_∞ norms of the computational error in solving the electric potential φ. In the first figure, the method preserves a second order accuracy on the uniform mesh. Moreover, even if the meshes in the centre of the domain are

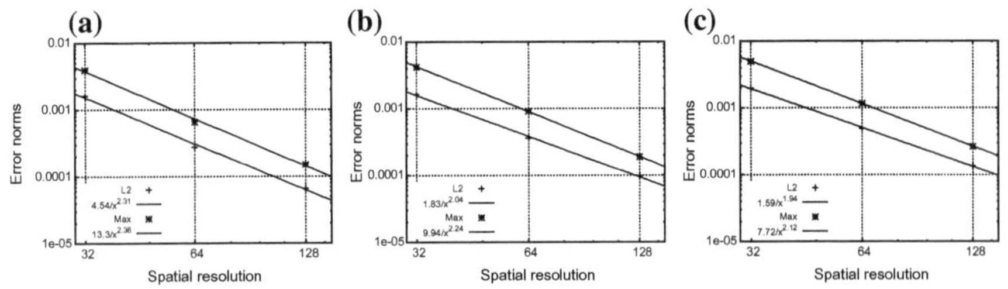

Fig. 4.1 Convergence studies in solving the electric potential by using the consistent scheme on the coarse/fine cell faces. **a** Uniform meshes; **b** One more level refinement in the center of the domain; **c** two more levels refinement in the center of the domain. As the results show, all the numerical solutions preserve the second-order accuracy

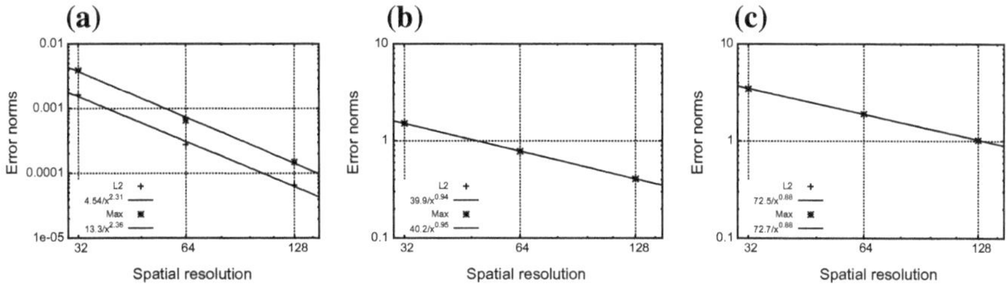

Fig. 4.2 Convergence studies in solving the electric potential by using the inconsistent scheme on the coarse/fine cell faces. The other descriptions are identical with Fig. 4.1. As it shows, although a second-order accuracy is still obtained on the uniform meshes, however, it will cause errors when the meshes are non-uniform

locally refined one or two extra levels, as shown in the second and third figures, the accuracy still converges to second-order. This test provide an evidence that *Scheme I* proposed in Sect. 3.1.1.1 is consistent on the non-uniform grid system.

Next, we use *Scheme II* to calculate $\boldsymbol{J}_{nf}$ on the cell faces, and the results are shown in Fig. 4.2. As it clearly depicts, if the meshes are uniform, a second-order convergence for φ is still obtained in the first figure. However, in the following two figures when additional one or two levels are added to the centre of domain, the accuracy will decrease to first-order. Otherwise, the error norms are inversely increased when the meshes are refined with extra levels in the centre of the domain, indicating the inconsistent treatment of $\boldsymbol{J}_{nf}$ will lead to numerical errors in estimating the flux of the current density on the coarse/fine cell faces.

4.1.2 MHD Flow in a Diverging Channel

To test the performance of the numerical algorithms in simulating MHD flows with solid boundaries, we select one test case proposed in some other publications [1, 16] and the geometrical definition of the problem is shown in Fig. 4.3. A diverting channel is constructed in a 4×1 domain and two opposite-facing plates of y_{bot} and y_{top} cut the domain to restrict the fluid flow between them. The geometry is given as

$$y_{bot} = \begin{cases} y_1, & if\ 0 \le x \le 1, \\ y_2 + 0.5(y_1 - y_2)\left(1 + cos\left(\frac{\pi}{2}(x-1)\right)\right), & if\ 1 < x < 3, \\ y_2, & if\ 3 \le x \le 4 \end{cases}$$

$$y_{top} = 1 - y_{bot}$$

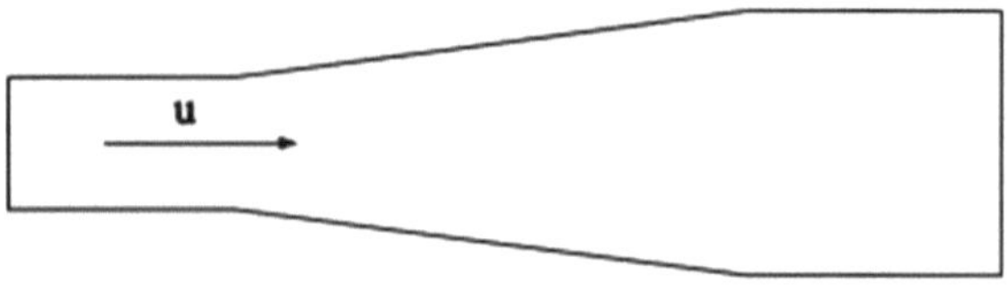

Fig. 4.3 The geometry of the diverging channel

A uniform velocity is imposed at the inlet while the convective boundary conditions are set at the outlet. Then we impose an external MF perpendicular to the flow field given as B_z, and the homogeneous Neumann condition for the electric potential is imposed on the solid boundaries.

Since the MF is along the z-direction, electric potential is generated along the solid surface. Due to the Ohms law of Eq. 2.11, when the flow recovers from the diverging region, the current density is zero and the flow should remain unaffected by the external MF. Correspondingly, analytic solutions of the velocity and the electric potential at outlet can be calculated.

All the numerical simulations are set with $CFL = 0.9$ and stopped at $t = 3.05$ by reaching a steady flow. The non-dimensional parameters are $Re = 10$ and $Ha = 10$ by using the half width of the inlet as the characteristic length. Firstly, three simulations are performed on uniform grids with $L = 5, 6$ and 7 levels of refinement. Then we investigate the variation of the error norms when the mixed-cells cut by solid boundaries are refined with additional levels. The solution of the outlet velocity is given as

$$u = -4.6875(y^2 - 0.16) \tag{4.4}$$

If we solve the electric potential Poisson equation based on this velocity distribution along the outlet, the solution of φ is

$$\varphi = -46.875\left(\frac{y^3}{3} - 0.16y\right) + c \tag{4.5}$$

Based on the uniform grids of $L = 6$, the numerical solutions of the velocity and the electric potential at the outlet are presented in Fig. 4.4. Both of them are in good agreement with the analytic solutions, indicating that the cut-cell method is accurate to be applied in solving the electrically insulating wall bounded MHD flows.

The error norms with different spatial resolutions are presented in Fig. 4.5 that the mixed-cells cut the solid boundaries are refined in an additional level. Both of L_1, L_2 norms indicate the preservation of a second-order of accuracy in space while the maximum error norms converge to first-order accuracy. It is proved that the numerical solutions are globally second-order accurate except in the local regions near the solid boundary.

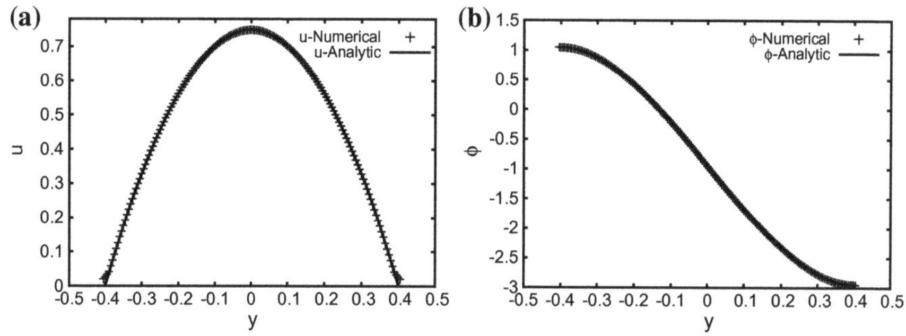

Fig. 4.4 Comparisons between the analytic solutions and the numerical results at the outlet of the diverging channel. **a** Velocity; **b** electric potential

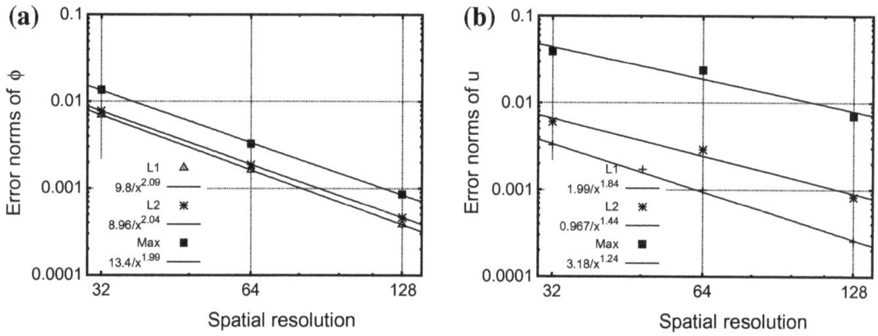

Fig. 4.5 The convergence order of the numerical results in calculating **a** electric potential and **b** velocity on non-uniform grids that the mixed-cells cut by solid boundaries are refined with one additional level

4.1.3 MHD Flow in Rectangular Channels

There are some exact solutions for fully developed incompressible laminar flows in 3D ducts with transverse MFs. Shercliff's case with four electrically insulating walls [19] and Hunt's case with two electrically conducting side walls and two electrically insulating Hartmann walls [6] are used to validate the consistent and conservative scheme developed in the present work.

4.1.3.1 Shercliffs Case

The Shercliff's case is simulated in a square channel whose walls are electrically insulating. The liquid metal flows in x-direction with a cross section of $2d \times 2d$ in y- and z-directions. A uniform MF is applied along the y-direction, and the schematic sketch of this problem, as well as the following Hunt's case, is shown in Fig. 4.6. The

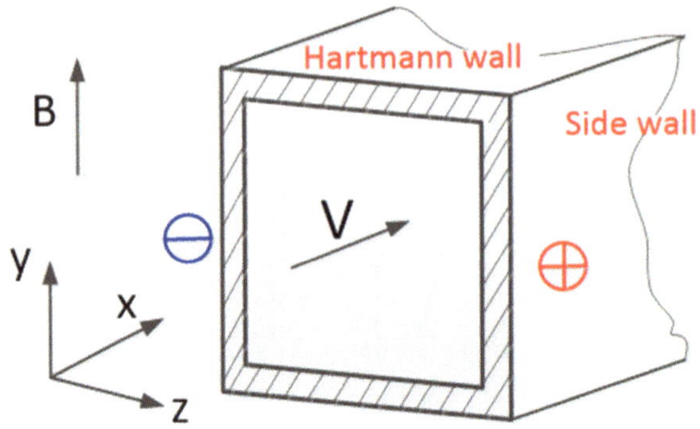

Fig. 4.6 The schematic sketch of the 3D duct flows under the influence of external MF

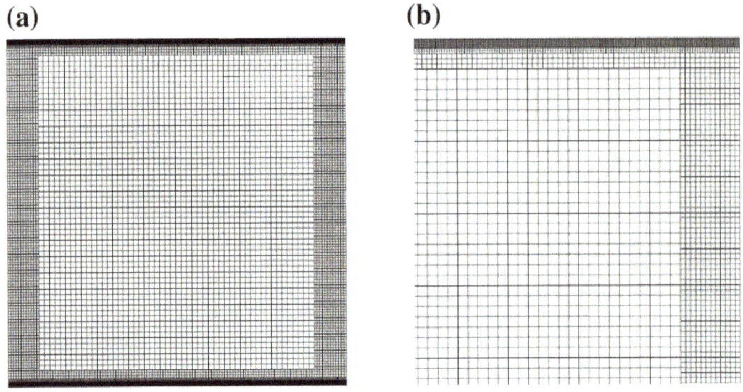

Fig. 4.7 The adaptive mesh distribution in the computational domain. **a** The overall view; **b** the local view

inflow velocity is selected as the characteristic velocity and d is the characteristic length, the Hartmann number is $Ha = 100$ in the present case.

Compared to the coarse meshes used in the duct center, finer grids with minimum sized of $1/3Ha$ are applied in the Hartmann layers while much finer grids with minimum sized of $1/5\sqrt{Ha}$ in the side layers, and the spatial resolution on the cut plane perpendicular to the flow direction is shown in Fig. 4.7.

The consistent and conservation scheme is used to calculate the fluxes of the current density on the coarse/fine cell faces, and the Lorentz forces are estimated based on the conservative formulation of Eq. 3.16. A comparison between the analytic solution and the numerical solution is presented in Fig. 4.8, and a good agreements are observed between them, indicating that the numerical methods are still accurate in the 3D simulations.

Furthermore, Fig. 4.9 gives the CPU time and the CPU time per mesh used in solving Shercliff'case on the uniform and adaptive non-uniform grids, respectively. A cluster-compatible Intel 2.3 GHz machine is used. From Fig. 4.9, a speedup of seven times is observed when using the non-uniform grids compared with the uniform grids.

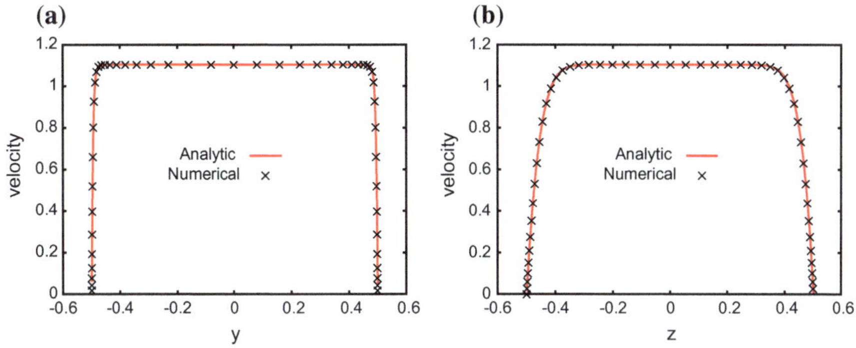

Fig. 4.8 Numerical results of Shercliff's case by using a consistent and conservative scheme based on the AMR framework, $Ha = 100$. **a** The comparison of the analytic solution against the numerical result along the centerline perpendicular to the Hartmann walls. **b** along the centerline perpendicular to the side walls

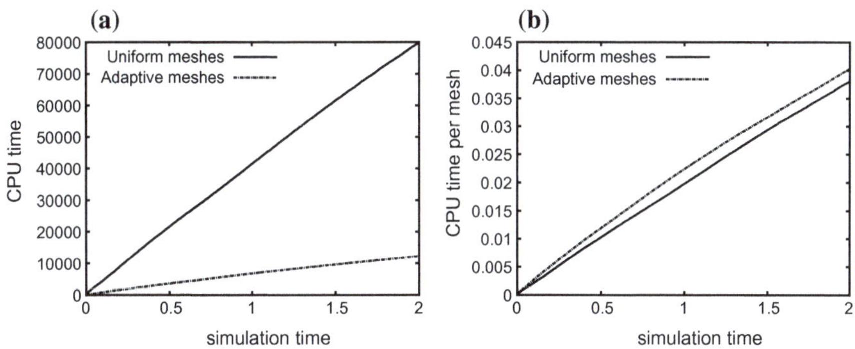

Fig. 4.9 Timings for uniform and nonuniform grids for the Shercliff's case. **a** The comparison of the total CPU time; **b** the comparison of the CPU time for per mesh

This demonstrates the efficient performance of the AMR technique in simulating the MHD flows. However, if the CPU time for per mesh is compared, as shown in Fig. 4.9, the nonuniform grids seem to be more inefficient. This is mainly because of the complex interpolation schemes in calculating the flux on the coarse/fine mesh faces when solving the pressure Poisson equations and the electric potential Poisson equations.

4.1.3.2 Hunt's Case

In this section, the fully developed Hunt's case in a square channel with $Ha = 100$ is numerically simulated. The Hartmann walls perpendicular to the external MF are assumed to be electrically conducting while the side walls parallel to the MF are electrically insulating. The consistent scheme developed in Sect. 3.2.2 for conjugate electromagnetic field calculation is used to treat the Hartmann walls. The Lorentz force is calculated based on the conservative formulation of Eq. 3.16, especially for those irregular mixed-cells.

We still adopt the non-uniform grids to conduct the numerical simulations. Nevertheless, the flow pattern in Hunt's case is significantly affected by the non-dimensional conductance ratio, which is given by

$$c_w = \frac{\sigma_w t_w}{\sigma_f d} \tag{4.6}$$

where the subscript w and f denotes the solid wall and fluid respectively, t_w is the thickness of the Harmann wall. Although c_w describes the ratio of the electrical conductivity between the wall and the fluid, however, there are three different cases according to the symmetry between the two Hartmann walls: (i) fully symmetrical case; (ii) the electrical conductivities are different between the two walls; (iii) the thickness are different between the two wall. The analytical solution of the symmetrical case is given by Hunt [6] and the later asymmetrical case is investigated by Tao and Ni [22].

For the symmetrical case with $c_w = 0.4$, the contour maps for the velocity distributions and the streamlines of the current density are shown in Fig. 4.10. The pictures illustrate that the Hartmann layers and the side layers are accurately resolved and the numerical results do not suffer from any numerical instabilities. Moreover, Fig. 4.11 presents the comparison between the analytical solution and the numerical results, and good agreements are observed between them.

Then an asymmetrical case is simulated that the conductivity of the top wall is twice the magnitude of the bottom wall. The velocity profile are presented in Fig. 4.12 that good agreements are observed with analytic solutions. Moreover, as Fig. 4.12a

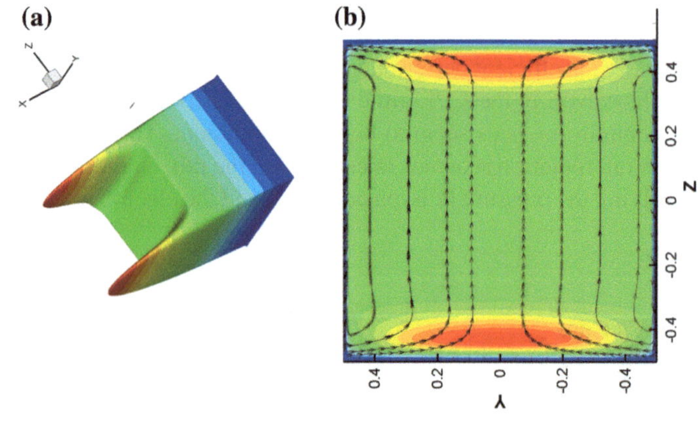

Fig. 4.10 Numerical results of fully developed Hunts case with symmetrical wall, Ha = 100. **a** The contour map of the streamwise velocity; **b** the streamlines of the current density on the plane perpendicular to the flow

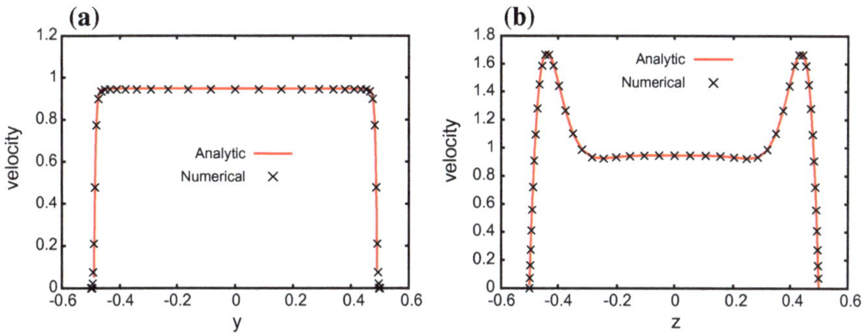

Fig. 4.11 Numerical results of Hunt's case with symmetrical Hartmann walls based on the consistent and conservative scheme within the AMR framework, $Ha = 100$ and $c_w = 0.4$. The other descriptions are as the same as those in Fig. 4.8

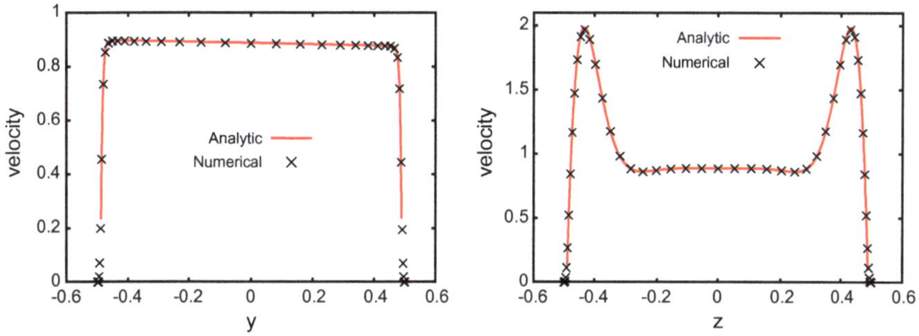

Fig. 4.12 Numerical results of Hunt's case with asymmetrical Hartmann walls that the conductivity of the top wall is twice the magnitude of the bottom wall, $Ha = 100$ and $c_w = 0.4$. The other descriptions are as the same as those in Fig. 4.8

shows, the velocity profile parallel to the magnetic field is not symmetrical any more, the wall with higher conductivity decrease the velocity more significantly.

Lastly, we simulate the asymmetrical case when the thickness of the bottom wall is twice the thickness of the top wall. The comparison between the numerical solution and the analytic solution is presented in Fig. 4.13, and the two results are observed to be consistent. Otherwise, due to the different thickness of the walls perpendicular to the MF, the velocity profile is also asymmetrical that the thicker wall produce larger velocity reduction.

The results presented above prove that the cut-cell approach together with the VOF method are sufficient to simulate the MHD flows with complex solid boundaries even if the solid is electrically conducting.

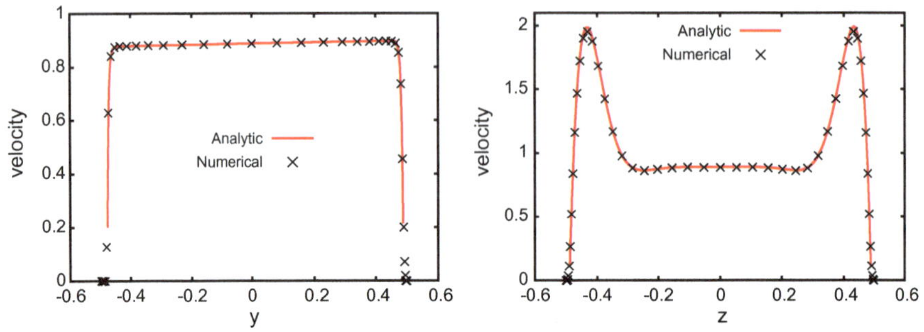

Fig. 4.13 Numerical results of Hunt's case with asymmetrical Hartmann walls that the thickness of the bottom wall is twice that of the top wall, $Ha = 100$ and $c_w = 0.4$. The other descriptions are as the same as those in Fig. 4.8

4.1.3.3 MHD Flow in Circular Pipes Having Finite Conductivity

In another type of circular pipe flow, the influences of the MHD effect are also investigated in case of electrically insulating walls and with electrically conducting walls, respectively [3, 18, 20, 21]. However, numerical difficulties are encountered in obtaining analytic solutions when Ha is large due to the infinite series expansions involving modified Bessel functions, and therefore an asymptotic method is developed under such circumstance. Other than that, Chang [3] studied the MHD flows in a circular pipe if the electrical conductivity of the wall was varied, no M type velocity profile appeared in his solution. Compared with Changs results, a later analytic work conduced by Samad [18] gave the M type velocity profile at low Ha case with highly electrically conducting wall.

In this section, we will numerically insulated the MHD flow in a circular pipe having finite conductivity and finite thickness under uniform transverse MF, and afterwards, the numerical results are compared with Samad's analytic solutions. Let the inner and outer radius of the pipe be R and $1.1R$, respectively, and moreover, the outer boundary of the pipe is also reconstructed by applying the VOF method. As aforementioned, all the available analytical solutions are limited within $Ha < 18$, however, more recently we can extend the parameter space to $Ha = 30$ due to the development of computers. Therefore, the numerical results respectively for $Ha = 18$ and $Ha = 30$ are compared with the analytical solutions, as shown in Fig. 4.14. It is worthy to notice that in both figures, c is the conductance ratio between the wall and the fluid, which is defined as $c = \frac{\sigma_w}{\sigma_f}$ that two values are considered in each Ha case, respectively of $c = 1$ and $c = 10$. As shown in Fig. 4.14, the M type velocity profile is more obvious with larger c, besides, all the numerical results agree well with the analytic solutions.

Furthermore, the numerical simulations are extended to a higher Ha case, i.e. $Ha = 100$ and the results are compared with Chang's asymptotic solution, as shown in Fig. 4.15. In the figure, c_w is defined as $c_w = \frac{\sigma_w d_w}{\sigma_f R}$ where d_w and R are the wall

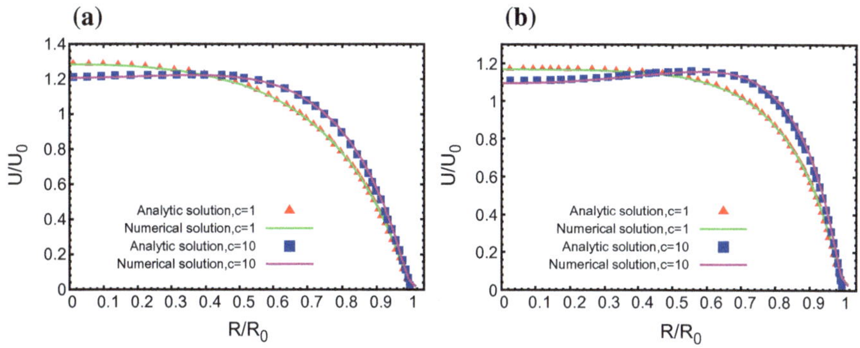

Fig. 4.14 Comparison of the outlet velocity profiles between numerical results and the analytic solutions at **a** $Ha = 18$ and **b** $Ha = 30$. Two electric conductivities are considered, respectively of $c = 1$ and $c = 10$

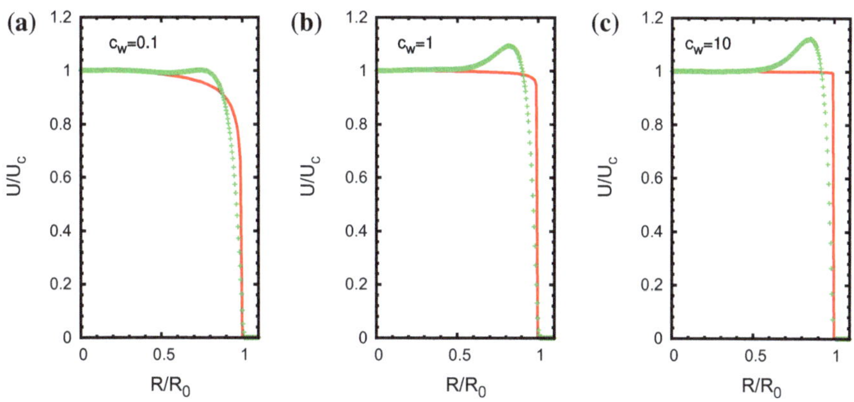

Fig. 4.15 Comparison of the outlet velocity profiles between numerical results (green points) and Changs asymptotic solutions (red lines) at $Ha = 100$, the wall conductance ratios are respectively of 0.1, 1, 10 from left to right. The results match well in the core region but overspeed zone appears near the wall in numerical results while the asymptotic approximation profiles do not show such flow behaviors

thickness and the pipe inner radius, respectively. Then $c_w = 0.1, 1, 10$ are subsequently selected to conduct the numerical simulations, nevertheless, there is a significant discrepancy by comparing with the asymptotic solutions: small zones of overspeed appear near the wall in the numerical solutions, while the profiles seem to be flat in asymptotic approximations. Moreover, the figure shows that the velocity jets appear more apparent when c_w increases. However, this M type velocity distribution is identical with Samad's analytic solutions at $Ha = 18, 30$.

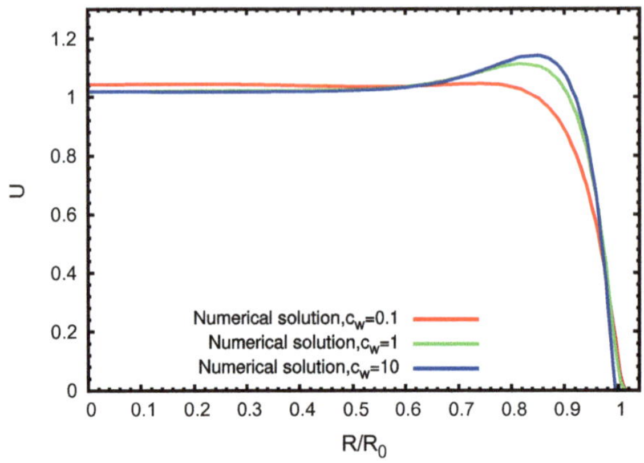

Fig. 4.16 Numerical results of the pipe flows with different wall-conductances, $Ha = 100$. It clearly shows that as Ha number increases, the M type profile is more obvious

Finally, the numerical results in case of $Ha = 100$ with different c_w are put together in Fig. 4.16.

4.1.3.4 MHD Flow Past a Circular Cylinder

At last, the 3D MHD flows past a circular cylinder, as extensively studied by Mutschke [14] and Grigoriadis [5], are lastly presented. As it is widely known, the transition scenario from 2D instability to 3D instability happens approximately at $Re = 194$ [24] in the ordinary flows. When the flow becomes 3D, the electric potential is generated in presence of an streamwise applied MF. Based on the Cartesian cut-cell approach, the numerical results are compared with the available publications [5, 14] when the strength of the MF is varied, respectively of $N = 0, 0.2, 1.0$.

The computational domain is schematically shown in Fig. 4.17, the cylinder is located at a distance of $L_1 = 8d$ from the inlet while the outlet is $L_2 = 24d$ downstream behind it, and d is the diameter of the cylinder. The duct walls and the cylinder are assumed to be electrically insulating. In the transverse z-direction, the cylinder is set at the center and we set $L_z = 16d$ in order to minimize the effect of the outer boundary on the development of the wake. The coming flow is set with an inflow velocity of u while the convective outflow condition is set at the outlet. Besides, the impermeable boundary conditions are applied on the other walls. A homogeneous MF is imposed along the streamwise direction, given as B_x.

To make the use of the AMR framework, two different grids are used, one is with the AMR technique (*Grids I*) while the other one without that (*Grids II*). The minimum size of the meshes around the cylinder is set with $\Delta x = 0.03125d$ in both grids, and such spatial resolution generates 32×32 cells to resolve the area of $d \times d$. Furthermore, within *Grids I*, the meshes are adapted every time step according

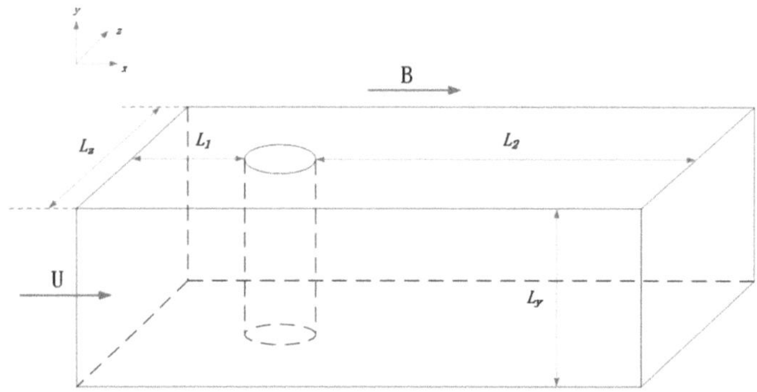

Fig. 4.17 The schematic sketch of the MHD flows past a circular cylinder

to the criterion of the vortex magnitude and the velocity gradient, with maximum size of $\Delta x = 0.25d$ and minimum size of $\Delta x = 0.03125d$. Based on such spatial resolutions, two Ha are considered, respectively of $Ha = 6.32$ and $Ha = 14.14$, which result in $N_x = 0.2$ and $N_x = 1.0$.

Throughout the whole simulations, the major parameters we study and compare are the lift coefficient C_l and the drag coefficient C_d, together with the recirculation pattern length X_r, and in addition, the Strouhal number St is used to describe the frequency of vortex generation when the flow becomes unsteady. All the above non-dimensional parameters are given as

$$C_l = \frac{F_l}{0.5\rho u^2 d}, \quad C_d = \frac{F_d}{0.5\rho u^2 d} \quad St = \frac{fd}{u} \tag{4.7}$$

where d and u are the characteristic length and velocity respectively, being the cylindrical diameter and the inflow velocity, respectively.

The time histories of C_d and C_l on both grids are presented in Fig. 4.18. Regarding the ordinary flow of $N_x = 0$, it gradually develops to a 3D unstable pattern as the C_l starts to oscillate. The flow transition from 2D to 3D is marked by a sudden decrease in the lift force as well as the drag forces, and this has also been studied by B. Mück et al. [13] for MHD flows around the rectangular obstacle. The St number in the 2D stage ($St = 0.210$) is slightly higher than the following 3D stage ($St = 0.197$), which is also found by B. Mück. Compared to the results on *Grids II*, the oscillations amplitude of C_d is more intense on *Grids I*, this is probably attributed to the dynamic distribution of the adaptive mesh in the computational domain.

In the second row, under a moderate MF of $N_x = 0.2$, the drag coefficient will drop sightly while a reduction in the lift coefficient is noticed, indicating that the flow is more stable compared to the ordinary flows. However, it still belongs to the 3D unstable flows being marked by the slight oscillations of C_l, this is also consistent with the numerical study by Mutschke [14] that the transition to a pure 2D flow is found to take place in the parameter space of $0.3 < N_x < 0.4$.

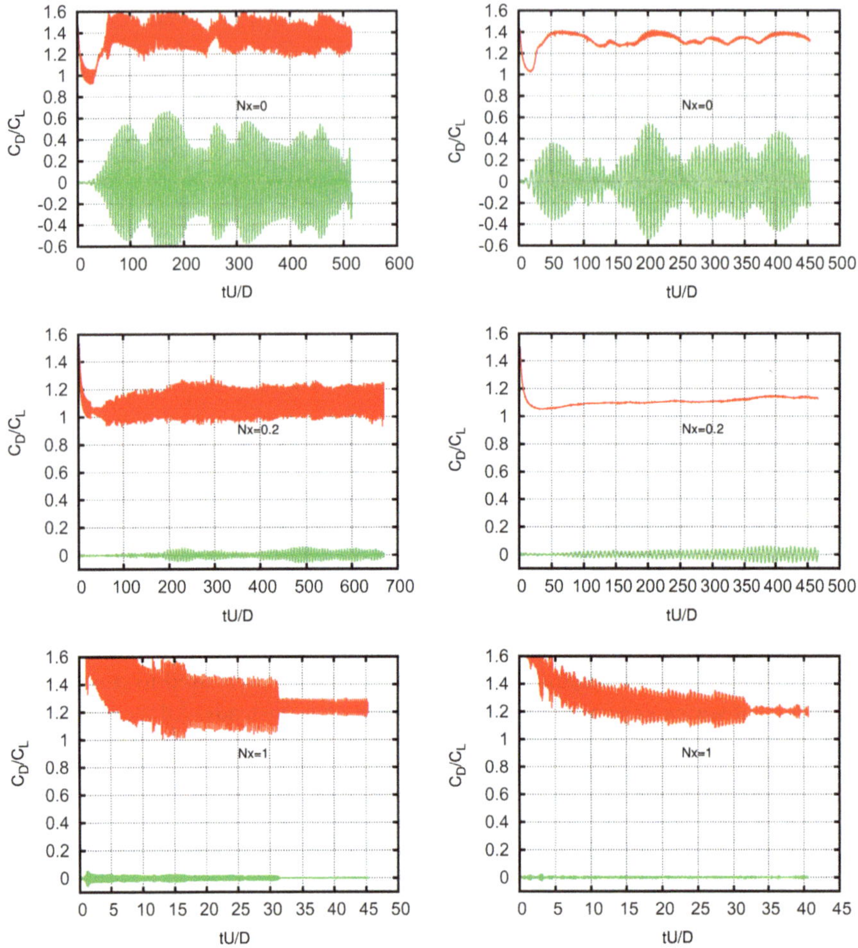

Fig. 4.18 The evolution of C_d (red line) and C_l (green line) under different MFs, respectively of $N = 0$, $N = 0.2$ and $N = 1.0$ from top to bottom. The left pictures are calculated on *Grids I*, while the right ones are on *Grids II*

In the last scenario by further increasing the MF to $N_x = 1$, the drag force coefficient is observed to rise in contrary, and this is also in agreement with some previous studies [5, 14]. In contrast, C_l is decreased to zero, indicating that the flow is more stable that the 3D instabilities are fully disappeared.

Table 4.1 presents all the numerical results in case of $Re = 200$. Both drag/lift force coefficients on different grids are found to be in good agreement with the available reference results.

Table 4.1 Comparisons of MHD flow past around a circular cylinder at $Re = 200$

Case (Grids)	N_x	C_d	C_l	X_r	St
Present (I)	0	1.33	0.63	–	0.197
Present (II)	0	1.32	0.57	–	0.196
Grigoriadis (coarse grid)	0	1.29	0.50	–	0.204
Grigoriadis (fine grid)	0	1.30	0.63	–	0.204
Mutschke	0	1.30	0.60	–	0.197
Present (I)	0.2	1.10	0.06	–	0.171
Present (II)	0.2	1.13	0.06	–	0.167
Grigoriadis (coarse grid)	0.2	1.04	0.07	–	0.174
Grigoriadis (fine grid)	0.2	1.08	0.09	–	0.172
Mutschke	0.2	1.07	0.09	–	–
Present (I)	1.0	1.23	0.0	3.17	–
Present (II)	1.0	1.20	0.0	3.18	–
Grigoriadis (coarse grid)	1.0	1.17	0.0	3.13	–
Grigoriadis (fine grid)	1.0	1.20	0.0	3.10	–
Mutschke	1.0	1.19	0.0	3.11	–

4.2 The Multi-Phase MHD Flows

Since either the implementation of the VOF method or the surface tension model has been validated by numerous tests in Gerris [17], therefore in the following study, we will focus to investigate the validity of the numerical algorithms in simulating the multi-phase MHD flows.

4.2.1 Single Gas Bubble Motion in the Liquid Under Magnetic Field

The behaviors of the gas bubbles rising in the liquid metal are firstly studied since the MFs are always used for bubble control in metallurgical process. The whole process is mainly described by the following non-dimensional parameters

- Eötvös number $Eo = \frac{\Delta\rho g d^2}{\sigma}$
- Morton number $Mo = \frac{\mu_1^4 \Delta\rho g}{\rho_1^2 \sigma^3}$

where the subscript '1' denotes the ambient phase, g the gravity and d the bubble diameter.

Otherwise, besides the Reynolds number Re, the Weber number We and the interaction number N as described aforementioned, another important non-dimensional parameter is the aspect ratio of $\chi = d_{max}/d_{min}$, where d_{max} is the longest axis and d_{min} is the shortest axis of the bubble after the shape deformation is stable.

4.2.1.1 Single Gas Bubble Motion Without Magnetic Field

Without MFs, we compare our numerical solutions with the available experimental results in a given parameter space, as presented by Bhaga and Weber [2]. The numerical parameters are chosen according to the experiments, and the numerical predictions of terminal Re are compared with the corresponding experimental results, as listed in Table 4.2 that good agreements are observed between them.

Furthermore, we take another numerical test according to the experimental study conducted by Zenit and Magnaudet [26] that a single bubble rises in the DMS-T02 silicone oil. The bubble diameter is $d = 1.48$ mm, resulting the terminal Re_T to be nearly 138 and the terminal aspect ratio of the bubble shape to be $\chi = 1.64$. The terminal bubble shape in our numerical simulation is shown in Fig. 4.19a with $\chi = 1.65$, which is very close to the experimental result, besides, the evolution of Re versus time is presented in Fig. 4.19b that the terminal Re approaches $Re_{ter} = 137$, which is also identical with the experimental result. Therefore, Gerris is proved to be accurate in simulating the single bubble rising in a viscous fluid.

Table 4.2 Comparisons of our numerical solutions with the available experimental results by Bhaga and Weber [2]

Group	Eo	Mo	Re_{exp}	Re_{num}	Difference (%)
A	116	5.51	13.3	13.4	0.75
B	116	0.103	42.2	41.9	0.71
C	115	4.63×10^{-3}	94	94.6	0.64

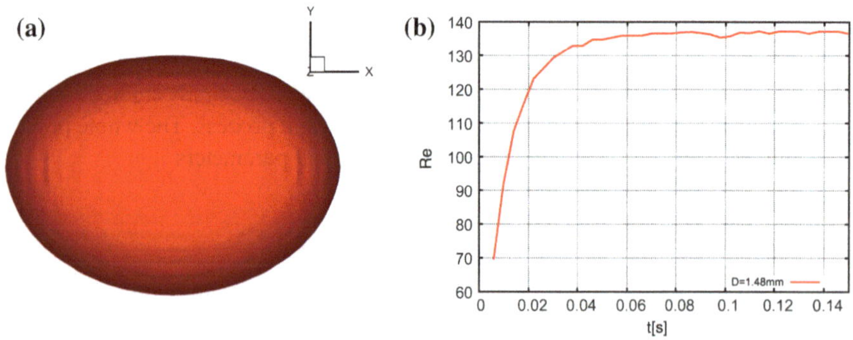

Fig. 4.19 A single bubble rising in the DMS-T02 silicone oil with an initial diameter of $d = 1.48$ mm. **a** The terminal bubble shape; **b** the evolution of the rising Re

4.2.1.2 Large Bubble Motion Affected by a Horizontal DC Magnetic Field

Recently, Zhang et al. [28] studied the bubble-driven flows under the influence of the horizontal MF by using the UDV measurement method experimentally. Miao et al. [11] performed the numerical simulations by using the business software of CFX, while a modified MHD turbulence model was developed. Firstly, We will investigate the behaviors of a large bubble motion under the influence of horizontal MF. Moreover, the numerical results are compared with the experimental observations and the numerical solutions respectively.

The simulations are conducted in a rectangular channel with no-slip walls, and the width of the channel is twice the diameter of the bubble. The sketch for this numerical case is presented in Fig. 4.20a, where the initial bubble position, the gravity direction and the magnetic direction are given. With a Eo number of $Eo = 24.5$, the terminal rising velocity results in $Re = 125$ and $We = 6.5$. The adaptive meshes are employed to capture the bubble interface accurately with a minimum size of $\Delta x = \frac{d}{64}$, which is shown in Fig. 4.20b that the mesh distributions in vicinity of the bubble are presented.

Besides, the top view of the bubble shape at $t = 1.08$ s under different N are displayed in Fig. 4.21, where the MF directs from left to right (x-direction). In Fig. 4.21a without the MF, the bubble shape is axi-symmetrical in the cross-section, however, it turns to be elliptical under the influence of a horizontal MF, given in Fig. 4.21b, c. The bubble is suppressed in the direction parallel to the MF, and this anisotropic effect is because of the asymmetric distribution of the Lorentz force.

Moreover, in case of $N = 100$, The streamlines of the current density on the cross-section beneath the bubble are displayed in Fig. 4.22, where the contour map

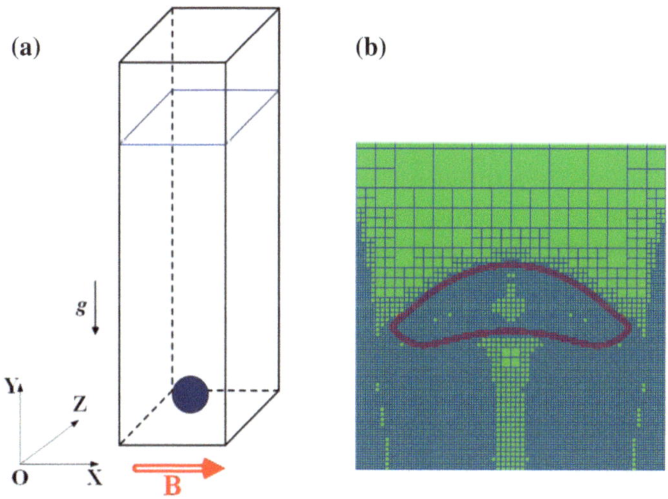

Fig. 4.20 The bubble motion under the influence of a horizontal MF. **a** The sketch of the problem; **b** the spatial resolution near the bubble

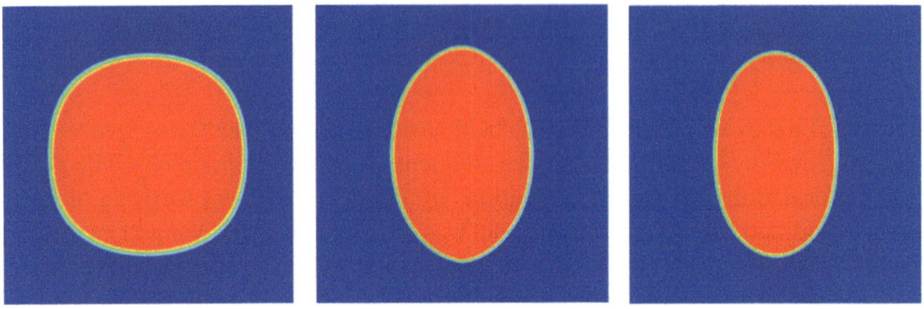

(a) $N = 0$ (b) $N = 25$ (c) $N = 100$

Fig. 4.21 The top view of the bubble shape **a** without magnetic field, **b** with horizontal MF of $N = 25$ and **c** $N = 100$ directing from left to right (x-direction)

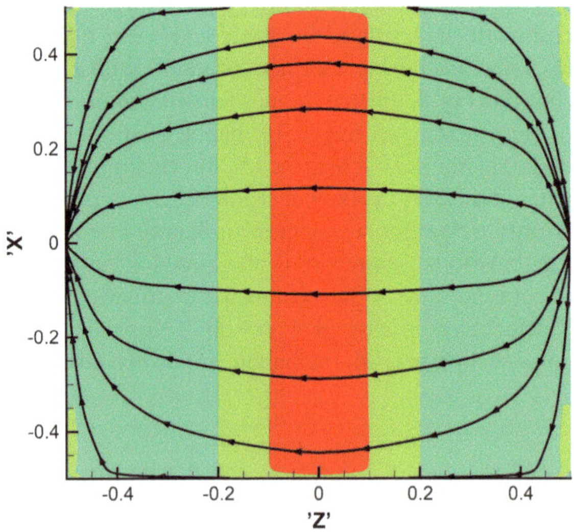

Fig. 4.22 The streamlines of the current density and the contour map of the vertical velocity on a cross section $4R$ beneath the bubble bottom. The dark color indicates the flows is downwards while the light color is upwards

represents the velocity distribution. Currents pass through the liquid metal and close in the Hartmann layer. However, the rising velocity is no longer homogeneous that the flow ascends in a narrow area along the MF (x-direction) and descends in the ambient region. This anisotropic phenomenon is also found by Miao et al. [11] in simulating the bubble driven flows.

To make such anisotropic distribution of the rising velocity be more clear, two vertical planes respectively parallel and perpendicular to the MF are cut out to show different structures of the flow field, as depicted in Fig. 4.23. Without MF as shown in Fig. 4.23a, b, the velocity distributions around the bubble are quite analogous in both orthogonal planes that the vortex structure are isotropic. However, the flow field becomes anisotropic when horizontal MF is applied, as shown in Fig. 4.23c, d. From the figures, the plane parallel to the magnetic field suffers a suppression in vortex field and the descending part of fluid disappears. On the contrary, the vortex field on

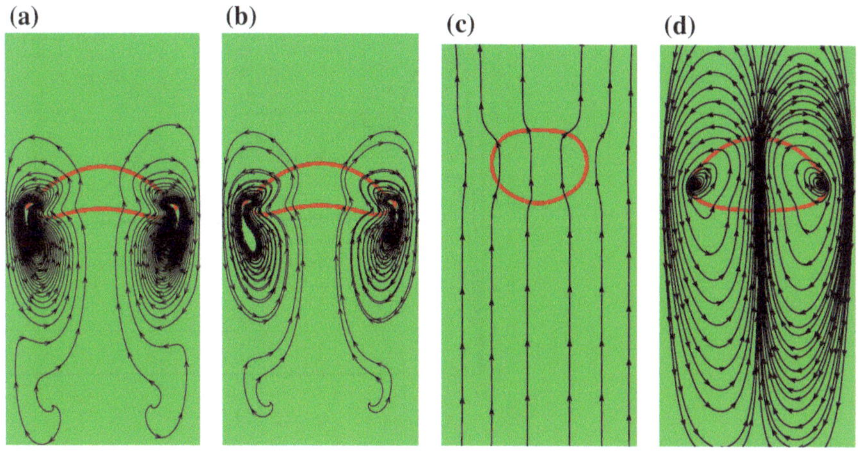

Fig. 4.23 Streamline of flow field **a**, **b** without and **c**, **d** with magnetic field. **c** is the plane parallel to the magnetic field while **d** is the plane perpendicular to it

the perpendicular plane is amplified and expanded to a larger region compared to the original flow. All the results are in good agreement with the available publications regardless of experiments [28] or numerical simulations [27], even if their studies are mainly bubble-driven flows. Moreover, the results displayed above are consistent with the theoretical predictions by Davidson [4] who has concluded that a vortex with its axis perpendicular to the MF will be damped, leading the velocity distribution independent of the coordinate parallel to the field direction.

Benefiting from the numerical simulations, we can provide more detailed information of the flow field than the experiment does. Figure 4.24 depicts the changing trend of the rising velocity versus time under the influence of different MFs. It shows the terminal velocity decreases monotonously with the magnetic intensity. This is accordant with the results given by Mori et al. [12] that the rising velocity of the bubble will decrease under stronger MF if the bubble rises rectilinearly.

4.2.1.3 Small N_2 Bubble Rising in Mercury Under a Horizontal Magnetic Field

In the above section, the large bubble rising in the liquid under the influence of horizontal MF is numerically simulated, and the numerical solutions are in qualitative accordance with the available experimental results. Furthermore, to be more quantitative, another group of numerical tests of small N_2 bubble rising in mercury under the influence of a horizontal MF is further simulated according to the available experiments conducted by Mori et al. [12], who employ an electrical triple probe to measure the velocity of the bubbles.

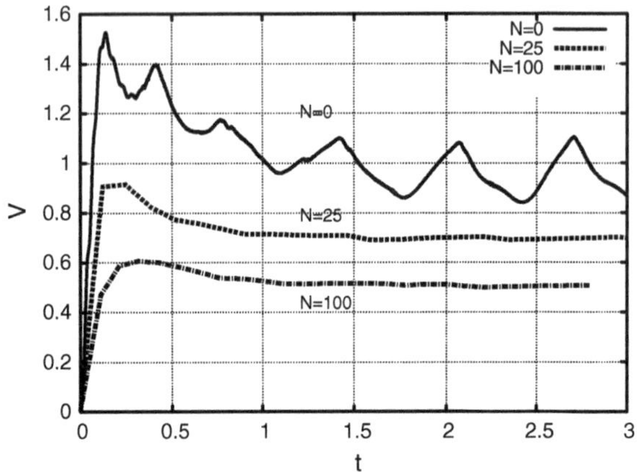

Fig. 4.24 Evolutions of the rising velocity under different horizontal MFs

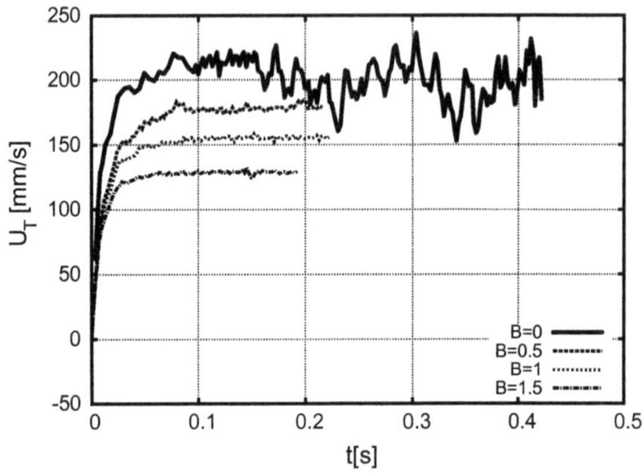

Fig. 4.25 The histories of the rising velocity of the N_2 bubble under the influence of different horizontal MFs

The N_2 bubble with radius of $r = 2.8\,\text{mm}$ is investigated in particular in our numerical simulation and the terminal Re_T approaches 3000 because of the large density difference between the liquid mercury and the gas. In addition, the MF is set with $B = 0$, $B = 0.5\,\text{T}$, $B = 1\,\text{T}$, and $B = 1.5\,\text{T}$ respectively, in order to be consistent with the experimental settings.

Time histories of the rising velocity of N_2 bubble under the influence of different horizontal MFs are presented in Fig. 4.25, and the terminal velocities are already stable when the computations are stopped. It is observed the terminal velocity to decrease with stronger MF as shown in the figure. Besides, without MF, the terminal velocity is found to oscillate around $u = 200\,\text{mm/s}$ with small fluctuations, which are gradually disappeared under $B = 0.5\,\text{T}$, $B = 1\,\text{T}$ and $B = 1.5\,\text{T}$.

Table 4.3 The terminal velocities of N_2 bubble rising in mercury under different horizontal MFs. Mori's [12] experimental results are also presented for comparison

Magnetic intensity (T)	0	0.5	1	1.5
Numerical (mm/s)	200	175	152	130
Experimental (mm/s)	201	168	143	130

Moreover, the comparisons between our numerical solutions and the experimental results are presented in Table 4.3. Good agreements between the terminal velocities are observed, whatever the strength of the MF is. It should be noted that without magnetic field, Mendelson [9] gives an analytical solution to predict the terminal velocity of small bubbles, and it turns to be $v = 198.77$ mm/s in this case. This theoretical prediction is very close to our numerical result.

Obviously, when the directions of the external MF are different, respectively parallel or perpendicular to gravity, different rising behaviors of the bubble motion would be expected, including the bubble shapes, the bubble rising paths and the vortex structures. Detailed investigations of the bubble motion in liquid metal under the influence of different MFs are discussed in the Chaps. 5 and 6 of this thesis.

4.2.2 *Thermocapillary Droplet Motion and Cavity Convection Without or with Magnetic Field*

In this section, the numerical simulations are performed to test the accuracy of the algorithm implemented in Sect. 3.2.2 for Marangoni effects.

4.2.2.1 The Thermocapillary Migration of a Droplet in the Limited of Zero Marangoni Number

The first numerical test is conducted to study the behavior of a droplet motion in an ambient liquid with linear temperature distribution. It is believed that the temperature gradient will drive the bubble moving towards the hot region and the flow characteristics are governed by several dimensionless parameters based on the properties of the ambient fluid, as shown in Sect. 2.2, and they are

$$Re = \frac{\rho a U_r}{\mu}, Ma = \frac{\rho c_p a U_r}{\lambda}, Ca = \frac{\mu U_r}{\sigma_0} \tag{4.8}$$

where a is the radius of the initial drop and U_r is the reference velocity defined as $U_r = \sigma_T \mid \nabla T \mid a/\mu$. ∇T is the temperature gradient in the ambient fluid. Then the dimensionless velocity $U^* = U/U_r$ and the reference timescale $t = a/U_r$ is used for the following simulations.

A 2D test case introduced by Nas and Tryggvason [15] is carried out firstly as a quantitative comparison. The simulation is performed in a square domain extending four drop radii in each direction, with the drop to be set in the center of domain. No-slip boundary conditions are imposed on the top and bottom walls, while the periodic boundary conditions are used in the horizontal directions. A linear temperature field is imposed along the vertical direction that the non-dimensional parameters are chosen as $Re = 2.5 \times 10^{-3}$, $Ma = 2.5 \times 10^{-3}$ and $Ca = 1.0 \times 10^{-3}$. The ratio of all the material properties between the drop and the ambient fluid are set to be 0.5, according to Nas and Tryggvason's simulation. Four different spatial resolutions are investigated to study the grid independence with 32×32, 64×64, 128×128 and adaptive meshes in domain respectively. They correspond to 16, 32, 64 grids per drop diameter in the first tree grids while in the last one, one more level refinement is added at the interface based on the 32×32 grid system.

Figure 4.26 shows the velocity vectors that a symmetrical vortex is clearly presented on this figure. After that, Table 4.4 presents the calculated terminal rising velocity on four grids respectively, while the corresponding results from Nas and Tryggvason [15] are list as comparisons. As the grids are refined, terminal velocities are observed to converge and the discrepancy between the last two grids is found to be less than 1.1%. Therefore, the 64×64 spatial resolution will be chosen in the following studies for accuracy consideration as well as saving computational resources. On the other hand, our solutions are in good agreement with the reference results, even when the AMR technique is applied.

In the limit of zero Ma and small Re, a fully 3D simulation of drop motion under Marangoni effect is further investigated and the analytic solutions are used for comparisons. For creeping flows, the terminal rise velocity of a spherical drop is derived by Young [25], given as

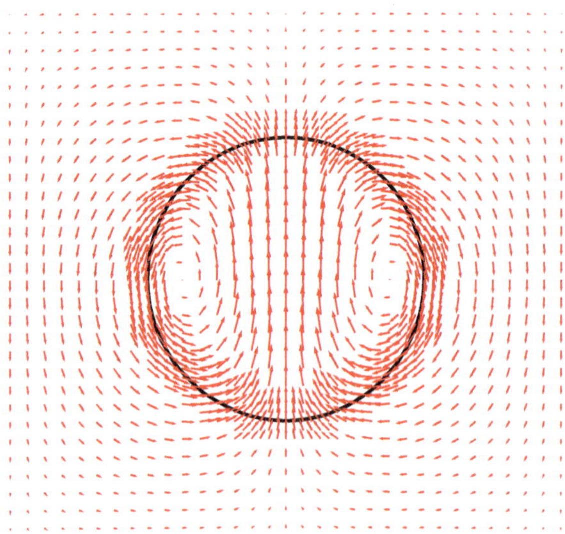

Fig. 4.26 Velocity vectors of the thermocapillary droplet motion

Table 4.4 Comparisons of the rise velocity on different grids. The reference results given by Nas and Tryggvason [15] are also presented for comparisons

Grid per droplet diameter	32	32/64	64	128
Present	0.1422	0.1464	0.1477	0.1493
NasTryggvason	0.134	–	0.145	0.149

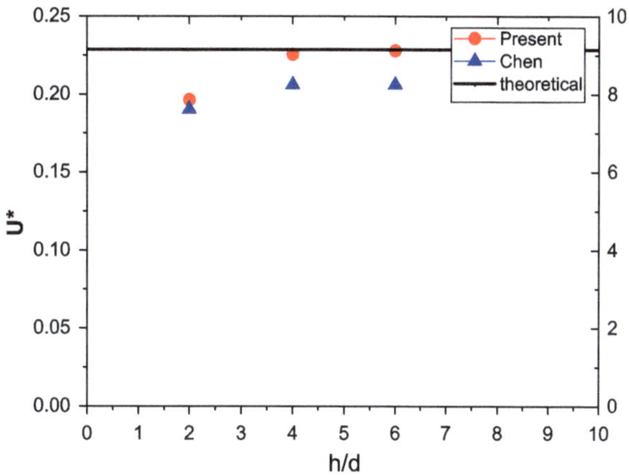

Fig. 4.27 The normalized rise velocity of a spherical drop versus the scaled distance. All simulations are carried out an a spatial resolution of 64 × 64 on the computational domain

$$U^* = \frac{2}{(2 + \lambda_1/\lambda_2)(2 + 3\mu_1/\mu_2)} \tag{4.9}$$

where the subscript 1 and 2 are the drop and the ambient fluid respectively. Figure 4.27 shows the terminal velocities versus different wall distances under parameters of $Re = 2.5 \times 10^{-3}$, $Ma = 2.5 \times 10^{-3}$, $Ca = 1.0 \times 10^{-3}$ and h denotes the distance between the walls while d is the drop diameter. When the distance between the spherical drop and the wall is larger than $2d$, the result is observed to converge to the analytic solution. Simultaneously, numerical results by Ma et al. [8] are also presented as comparison, and our results seem to be more accurate.

4.2.2.2 Thermocapillary Convection Without and with MF

The above simulations mainly concern about the droplet motion driven by Marangoni stresses and the numerical methodology developed in Sect. 3.2.2 is already validated. After that, we will further study the thermocapillary flow with the temperature gradient tangential to the fluid interface. The numerical model used for the present study is schematically depicted in Fig. 4.28a whereas a $4h \times 1h \times 1h$ cavity is filled with liquid that the air contacts with the free surface on top. The temperature is fixed at the left and the right walls respectively of T_1 and T_2 ($T_1 < T_2$). While the bottom wall is set with no-slip boundary condition, all other walls are set to be free-slip.

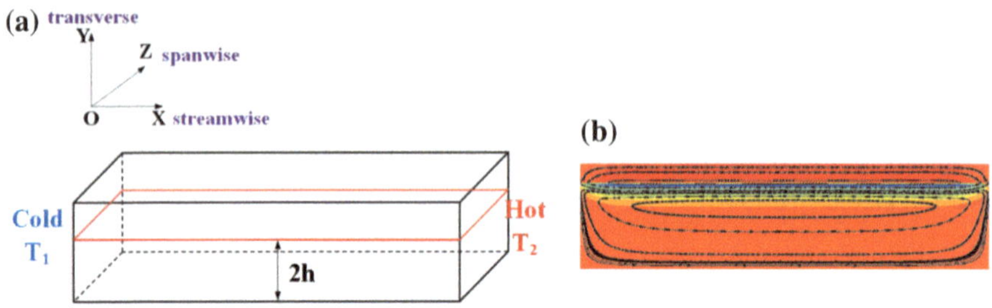

Fig. 4.28 Sketch of the thermocapillary cavity flows with horizontal temperature gradient. **a** the numerical model; **b** the thermocapillary driven flows under the Marangoni effect

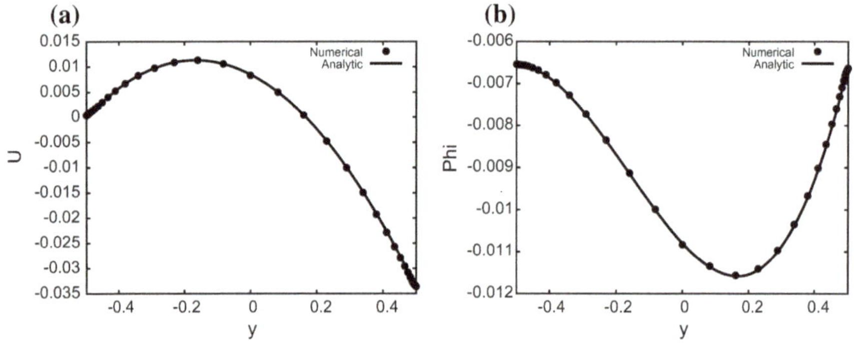

Fig. 4.29 The comparisons between the numerical results and the analytical solutions along the vertical line in the center. **a** The horizontal velocity; **b** the electric potential

Figure 4.28b shows the streamlines of the velocity in a cut plane perpendicular to z-direction when the horizontal MF is not applied, and an corresponding analytical solution is given by Levich [7] and Velarde [23] on the velocity distribution. From the top to the bottom, the analytical solution is in form of

$$U = \frac{\partial T}{\partial x}\frac{\partial \sigma}{\partial T}\frac{h}{8\mu}(3y^2 + 2y - 1) \quad (4.10)$$

where h is the half thickness of the liquid layer and y is nondimensionalized by h to fall in the range of $-0.5 < y < 0.5$. Based on this, the comparisons between the numerical results and the analytical solutions are presented in Fig. 4.29a whereas the x- axis is the nondimensional distance to the center of the layer, and a good agreement is observed between them. Furthermore, if an external MF is applied along the spanwise direction without calculating the Lorentz force, the induced electric potential is calculated by solving the electric potential Poisson equation. As a consequence, the comparison between the numerical result and the analytic solution is also presented in Fig. 4.29b, which shows a good consistency.

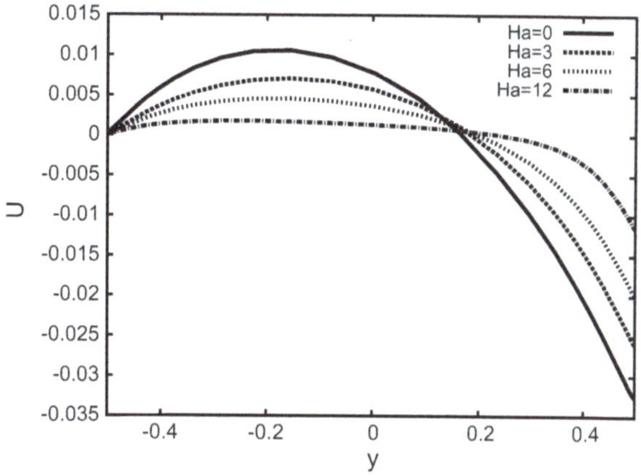

Fig. 4.30 The influence of the magnetic field on horizontal velocity distribution along the center vertical line

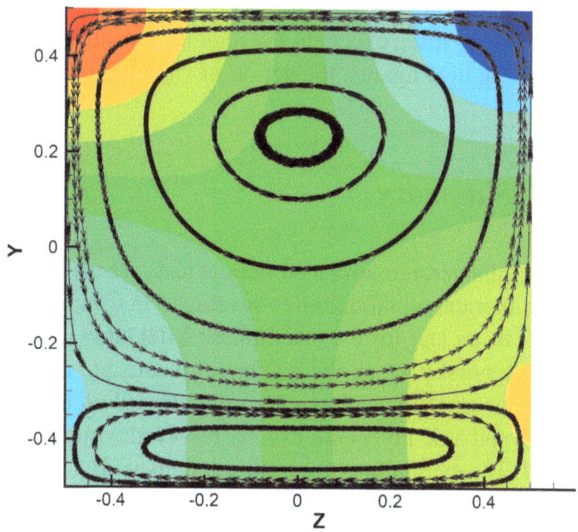

Fig. 4.31 The streamlines of the current density and the contour maps of the electric potential in case of $Ha = 3$ on the cut-plane

Furthermore, horizontal MFs are applied to investigate their influence on the thermocapillary convection. The variations of the horizontal velocity along the center vertical line are displayed in Fig. 4.30, whereas the magnetic strength is set as $Ha = 3$, $Ha = 6$ and $Ha = 12$ respectively. It is observed that the thermocapillary flow has been significantly suppressed by the MF. The horizontal velocity seems to be more homogeneously along the field direction under stronger MF.

A plane perpendicular to the streamwise direction is cut out at the center of the cavity, as shown in Fig. 4.31 that the current density streamlines are plot and the contour map represents the distributions of the electric potential, the MF is set with $Ha = 3$. It is observed that the $\nabla\varphi$ is in opposite directions at the top and bottom of the cavity and this is resulted from the back-flow. In vicinity of the free surface where the largest flow velocity appears, the induced electric field is more significant.

In the figure, the closed electric current loops are very similar with that of the Shercliff's case, and the directions of the current density are determined by the relative magnitude between $-\nabla\varphi$ and $\boldsymbol{u} \times \boldsymbol{B}$ as Eq. 2.11 shows. However, the physical mechanism hidden behind is different from the Shercliff's case because the flow changes its direction in the upper and lower part of the cavity. The current loop near the bottom wall is generated because of the boundary layer while it is more complicated for the formation of the current loop at top, that is because there is no boundary layer at the free surface. As shown in Fig. 4.28a, the horizontal velocity changes its direction at $y = 0.2$. In the region where $y \gg 0.2$, $(\boldsymbol{u} \times \boldsymbol{B})_z$ is significantly larger than that of $-\nabla_z\varphi$ and thus j_z directs towards $z-$; when y approaches $y = 0.2$, $(\boldsymbol{u} \times \boldsymbol{B})_z$ reduces to a value smaller than $-\nabla_z\varphi$ and therefore the current direction is changed towards $z+$; further in the region $y \ll 0.2$ where the flow direction has already changed, $(\boldsymbol{u} \times \boldsymbol{B})_z$ increases again and larger than $-\nabla_z\varphi$ that the current still directs towards $z+$.

Furthermore, the induced Lorentz force almost tends to suppress the flow and acts as a braking force except in the region near the bottom wall, where the Lorentz force drives the flow against the action of vicious braking.

4.3 Summary

In this chapter, numerous numerical tests are carried out to investigate the validity of the numerical methodologies developed in Chap. 3, respectively for the single-phase MHD flows and for the multi-phase MHD flows.

In the first part, all the numerical results are in good agreement with the analytic solutions and other reference results. Based on the AMR framework, the Hartmann layers and side layers are exactly resolved while the charge conservation law is satisfied. The estimation of the current density flux are consistent across the coarse/fine cell faces by using the present algorithms, however, an error occurs if an inconsistent scheme is applied. When the complex solid boundaries are electrically insulating, a cut-cell method is employed to compute the electromagnetic field on the Cartesian grids, the computational efficiency is observed to be improve by keeping the second-order accuracy in the bulk domain. In addition, if the solid boundaries are electrically conducting, a combined method coupling the VOF scheme and the cut-cell scheme is used to capture the jump condition of the electric conductivities in the mixed-cells cut by the solid boundaries, and the results are still accurate.

In the second part, the single bubble motion under the influence of horizontal MF is studied, and some similar results are obtained compared with the experimental results and numerical solutions. For the large bubble motion, the bubble shape and the vortex structure are observed to be anisotropic that the flow field is more independent of the direction along the field. In addition, the rising velocity of the bubble is always suppressed by the external MF. For the small N_2 bubble rising in the mercury, it is proved that the algorithms are still stable even if the multi-phase MHD flows are with great density ratio and electrical conductivity ratio between the fluids. The terminal rising velocity of the bubble is in good agreement with the experimental measurements. Furthermore, the vortex structures behind the bubble are also presented under different MFs, showing consistent performances with the theoretic predictions. The thermocapillary migration of a droplet is simulated numerically to test the correctness of the Morongoni stresses, and good results are obtained by comparing the present solutions with available theoretical and numerical solutions. Then the numerical model is applied to simulate the thermocapillary cavity flow and the results are still promising. Simultaneously, the electric potential is accurately calculated when external MF is imposed. Upon on this, we further investigate the influence of the MF on the thermocapillary cavity flow. It is observed the velocity distribution to be more homogeneous along the field direction.

References

1. Almgren AS, Bell JB, Colella P et al (1998) A conservative adaptive projection method for the variable density incompressible Navier-Stokes equations. J Comput Phys 142(1):1–46
2. Bhaga D, Weber ME (1981) Bubbles in viscous liquids: shapes, wakes and velocities. J Fluid Mech 105:61–85
3. Chang CC, Lundgren TS (1961) Duct flow in magnetohydrodynamics. Zeitschrift fr Angewandte Mathematik und Physik (ZAMP) 12(2):100–114
4. Davidson PA (1995) Magnetic damping of jets and vortices. J Fluid Mech 299:153–186
5. Grigoriadis DGE, Kassinos SC, Votyakov EV (2009) Immersed boundary method for the MHD flows of liquid metals. J Comput Phys 228(3):903–920
6. Hunt JCR (1965) Magnetohydrodynamic flow in rectangular ducts. J Fluid Mech 21(4):577–590
7. Levich VG (1962) Physicochemical hydrodynamics. Prentice Hall, Upper Saddle River
8. Ma C, Bothe D (2011) Direct numerical simulation of thermocapillary flow based on the volume of fluid method. Int J Multiph Flow 37(9):1045–1058
9. Mendelson HD (1967) The prediction of bubble terminal velocities from wave theory. AIChE J 13(2):250–253
10. Minion ML (1996) A projection method for locally refined grids. J Comput Phys 127(1):158–178
11. Miao X, Lucas D, Ren Z et al (2013) Numerical modeling of bubble-driven liquid metal flows with external static magnetic field. Int J Multiph Flow 48:32–45
12. Mori Y, Hijikata K, Kuriyama I (1977) Experimental study of bubble motion in mercury with and without a magnetic field. J Heat Transf 99(3):404–410
13. Mück B, Gnther C, Müller U et al (2000) Three-dimensional MHD flows in rectangular ducts with internal obstacles. J Fluid Mech 418:265–295
14. Mutschke G, Gerbeth G, Shatrov V et al (2001) The scenario of three-dimensional instabilities of the cylinder wake in an external magnetic field: a linear stability analysis. Phys Fluids 13(3):723–734
15. Nas S, Tryggvason G (2003) Thermocapillary interaction of two bubbles or drops. Int J Multiph Flow 29(7):1117–1135
16. Popinet S (2003) Gerris: a tree-based adaptive solver for the incompressible Euler equations in complex geometries. J Comput Phys 190(2):572–600
17. Popinet S (2009) An accurate adaptive solver for surface-tension-driven interfacial flows. J Comput Phys 228(16):5838–5866
18. Samad SKA (1981) The flow of conducting fluids through circular pipes having finite conductivity and finite thickness under uniform transverse magnetic fields. Int J Eng Sci 19(9):1221–1232

19. Shercliff JA (1953) Steady motion of conducting fluids in pipes under transverse magnetic fields. Math Proc Camb Philos Soc 49(1):136–144 (Cambridge University Press)
20. Shercliff JA (1956) The flow of conducting fluids in circular pipes under transverse magnetic fields. J Fluid Mech 1(6):644–666
21. Shercilff JA (1962) Magnetohydrodynamic pipe flow. Part 2:513
22. Tao Z, Ni MJ (2015) Analytical solutions for MHD flow at a rectangular duct with unsymmetrical walls of arbitrary conductivity. Sci China Phys Mech Astron 58(2):1–18
23. Verlarde MG (2012) Physicochemical hydrodynamics: interfacial phenomena. Springer Science and Business Media, Berlin
24. Williamson CHK (1996) Vortex dynamics in the cylinder wake. Ann Rev Fluid Mech 28(1):477–539
25. Young NO, Goldstein JS, Block MJ (1959) The motion of bubbles in a vertical temperature gradient. J Fluid Mech 6(3):350–356
26. Zenit R, Magnaudet J (2008) Path instability of rising spheroidal air bubbles: a shape-controlled process. Phys Fluids 20(6):061702
27. Zhang C, Eckert S, Gerbeth G (2006) Determination of the flow structure in bubble-driven liquid metal flows using ultrasound Doppler method. In: 5th international symposium on ultrasonic Doppler methods for fluid mechanics and fluid engineering, Zrich, Switzerland
28. Zhang C, Eckert S, Gerbeth G (2007) The flow structure of a bubble-driven liquid-metal jet in a horizontal magnetic field. J Fluid Mech 575:57–82

Chapter 5
The Argon Bubble Rising in the Liquid GaInSn Under the Influence of a Vertical Magnetic Field

As introduced in Chap. 1, the bubbly driven flows are encountered frequently in the metallurgical engineering and the casting engineering, where the bubbles are injected into the molten metal to stir and homogenize the liquid metal [1]. For such type of two-phase flows, the MFs are used to control the bubble motion through a contactless way. In addition, in the fusion reactor, the MHD effect will greatly suppress the turbulence of the flow in the energy transfer section, and a possible remedy is to inject gas bubbles into the coolant to enhance the flow disturbances. From this perspective, the bubble behaviors in the fluids subjected to the external MF has became a topic of considerable interest.

5.1 Introduction

During the last decade, the rise of a gas bubble in water has been the subject of many theoretical and experimental studies. It is now established that above a certain value of the Reynolds number ($Re > 600$) the bubble trajectory is no longer straight but exhibits zigzag oscillations or spiral motions [12]. The dynamics of this instability has therefore been the subject of detailed investigations. Saffman [19], Hartunian and Sears [10] and later Benjamin [2] tried to explain the instability mechanisms and the bubble shape on the basis of an assumed curl-free flow. Although the Saffman suggestion that the instability should be coupled with the wake oscillation is generally accepted, the predicted results do not agree quantitatively with the experiments.

Recently, thanks to the development of improved measuring techniques, such as the Dye visualizations [11], Laser Doppler Anemometry [7] and Particle Image Velocimetry [3], still in transparent liquids, it has been confirmed that the bubble path is strongly dependent of the wake instability, when the vorticity production around the bubble exceeds some threshold. Moreover, experiments performed in

J. Zhang, *The Developments and the Applications of the Numerical Algorithms in Simulating the Incompressible Magnetohydrodynamics with Complex Boundaries and Free Surfaces*, Springer Theses, https://doi.org/10.1007/978-981-10-6340-4_5

pure water and in tap water [6, 23] have demonstrated that the presence of pollutants could generate different rising behaviors and different shape oscillations. In tap water, these pollutants concentrate on the bubble deformable surface, modifying the surface tension and tending to make it be similar to a solid boundary with a no-slip requirement.

The numerical investigation of the bubble rising at high *Re* is particularly difficult, both because the continuous deformation of the bubble interface is coupled with the pressure fluctuations by surface tension, and because the fluid physical properties vary steeply across the interface. Mougin and Magnaudet [16] have numerically observed and confirmed the coupling of the path instability with the bubble wake behavior in the hypothetical case of a rigid ellipsoidal shape. A new numerical simulation by Gauditz and Adams [8] recently shows that, when the bubble motion makes zigzags, the unsteadiness is related to four hairpin vortices present in the wake, while a twisting vortex chain shows up when the spiral motion is present.

The case of a gas bubble rising in a liquid metal in the presence of an external MF is more difficult to be apprehended. The small value of the kinematic viscosity makes the Reynolds number be much larger than that in water, and the experimental measurements are more difficult because the liquid is opaque. This explains the scarcity of the published results in this research community, except for the experiment of a nitrogen bubble rising in mercury conducted by Mori et al. [15], and the more recent results given by Zhang et al. [25] about single argon bubble rising in liquid GaInSn, either under the influence of a horizontal or a vertical MF. These experiments reveal that the MHD effect is quite subtle, since small bubbles are accelerated by moderate MF, whereas large bubbles are retarded.

Numerical difficulties also come from the presence of the external MF, because the computations of the electric potential and the current density require an accurate scheme to couple with the fluid flow [17, 18]. Although some other attempts [9, 14] have been made, either by using the business software CFX with a modified turbulence model, or with a hybrid particle-level-set method, however, few general conclusions can be derived from these approaches.

Therefore, in this chapter, we will use the numerical methods proposed in Chap. 3 to simulate the single Ar bubble rising in the liquid GaInSn without or with vertical

Table 5.1 The physical properties and the corresponding non-dimensional parameters in the numerical simulations

	Ar	GaInSn
Density ρ(kg/m^3)	1.654	6361
Dynamics viscosity μ(kg/ms)	1.176×10^{-5}	2.2×10^{-3}
Electric conductivity $\sigma_e(\Omega^{-1}\mathrm{m}^{-1})$	0.1	3.27×10^{6}
Surface tension coefficient σ(N/m)		0.553
Morton number Mo		2.4×10^{-13}
Eötvös number Eo		0.74–4.9
Reynolds number Re		2000–4000
Weber number We		2–5

MF, and subsequently, the influence of the horizontal MF on the single bubble motion will be discussed in the next chapter. It should be noted that for the convenience of comparisons, all the parameters used here are set according to the experiments conducted by C. Zhang et al. [25], and the physical properties of the liquids are given in Table 5.1, in which the dimensionless numbers are also presented. However, we should keep in mind that for the measurement requirements, the experimental liquids are mixed with oxides so that the GaInSn are no longer pure, as described in Chap. 1.

5.2 The Bubble Motion Without MF

The diameters of the argon bubble are set between 2.5 mm and 6.5 mm, which result the dimensionless parameters to vary in the range of $Eo = 0.74$–4.9 and $Re = 2000$–4000. In addition, the ratios of the physical properties between the liquid GaInSn and Ar are very large, being almost $\rho_1/\rho_2 = 3846$, $\mu_1/\mu_2 = 187$ and $\sigma_{e1}/\sigma_{e2} = 3.27 \times 10^7$, respectively. In the experiments, the maximum vertical MF is set with $B = 0.3T$ and this will be extended to $B = 0.5T$ in the numerical simulations to provide more results. The computational domain is set with the size of $L_x \times L_y \times L_z = 20d \times 40d \times 20d$, and it is sufficiently large to ignore the wall effect. Bubbles are released at the bottom of the domain with an initially spheroidal shape and it then rises under gravitational force.

To be first, we will investigate the independence of the spatial resolution in simulating such problems. the results in case of $Eo = 2.2$ are shown in Fig. 5.1, where $Level-1$ to $Level-3$ indicate the smallest size of the grid in vicinity of the interface to be $\Delta = \frac{d}{25}$, $\Delta = \frac{d}{50}$ and $\Delta = \frac{d}{100}$, respectively. Besides, as described in Fig. 5.1a, the rising heights of the bubble centroid under different spatial resolutions are compared before the path instability happens. An obvious difference is

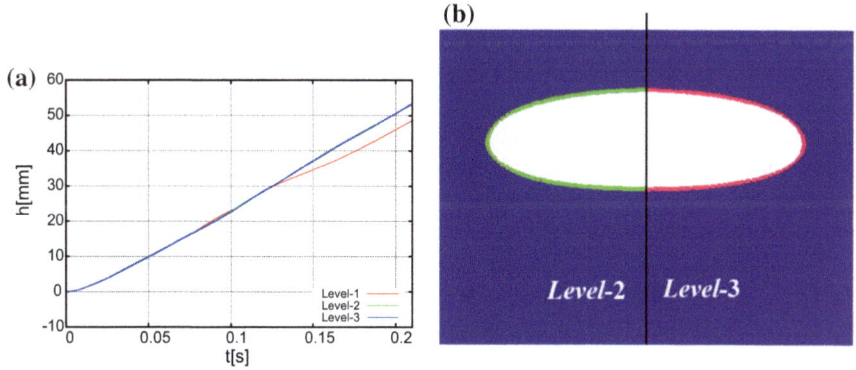

Fig. 5.1 The numerical results of a singe bubble motion in case of $Eo = 2.2$ and $N = 0$, they are obtained on different spatial resolutions of $Level-1$, $Level-2$ and $Level-3$, respectively. **a** Rising heights over time; **b** the bubble shapes

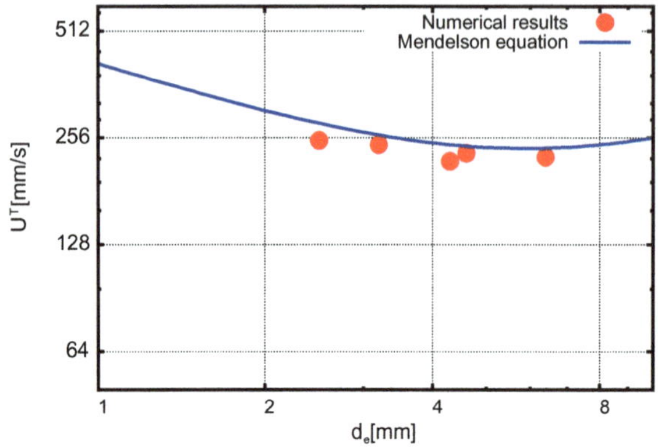

Fig. 5.2 Terminal rising velocities of different sized bubbles rising in GaInSn. The solid line is the analytical solution of the Mendelson equation [13]

Table 5.2 The behaviors of the rising bubble without MF

No.	Eo	χ	Re	path
(a)	0.74	2.01	1843	Oblique
(b)	1.21	2.12	2313	Spiral
(c)	2.2	2.41	2797	Spiral-like
(d)	4.9	2.54	3792	Chos

observed between the results of $Level - 1$ and the other two grids, while the results obtained from $Level - 2$ and $Level - 3$ are almost identical. In Fig. 5.1b, the bubble shapes projected onto the center-vertical plane are compared between $Level - 2$ and $Level - 3$ at $t = 0.05$ s. The bubble shapes based on the two spatial resolutions are in good agreement, and therefore, $Level - 2$ together with the AMR technique applied in the vortex region is selected for the numerical simulations.

In Fig. 5.2, the terminal velocities versus the diameters of the rising bubble without MF are presented, moreover, the theoretical solutions of the Mendelson equation [13] are also attached for comparison, and good agreement is observed between them.

Besides the terminal velocities, some other rising behaviors of the four representative bubbles are presented in Table 5.2 without MF. In the table, χ represents the aspect ratio between the major and minor axes of the bubble. It is observed that with larger bubble size, the bubble deformations are increased and the rising paths are more complicated.

Furthermore, Fig. 5.3 shows the top view of the bubble paths collected from Table 5.2. In Fig. 5.3a where the bubble size is the smallest, an oblique trajectory is observed with little oscillations. Both Fig. 5.3b, c show spiral rising paths, and the first one is more regular while the second one is more likely to be flattened. Finally, the largest bubble has a random, rocking motion and therefore the chaos rising path is observed, as shown in Fig. 5.3d.

Afterwards, the streamwise vortex structures behind the bubbles are plotted in Fig. 5.4, where the green color represents $\omega_y = 100$ and the blue color is $\omega_y = -100$,

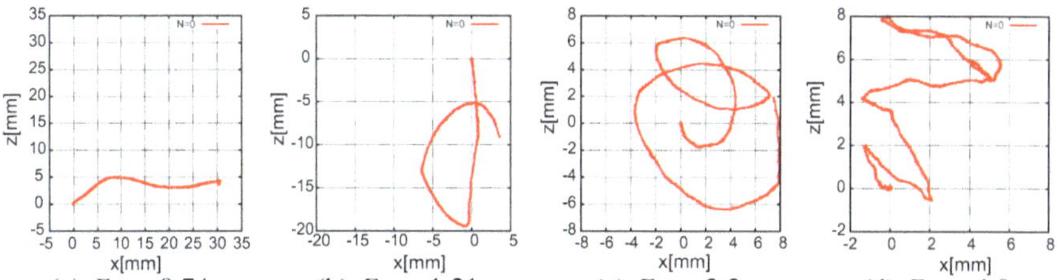

(a) $Eo = 0.74$ (b) $Eo = 1.21$ (c) $Eo = 2.2$ (d) $Eo = 4.9$

Fig. 5.3 The top view of the rising paths of different sized Ar bubbles, from $Eo = 0.74$ to $Eo = 4.9$. It is observed that the rising paths are more complicated as the bubble size is increased

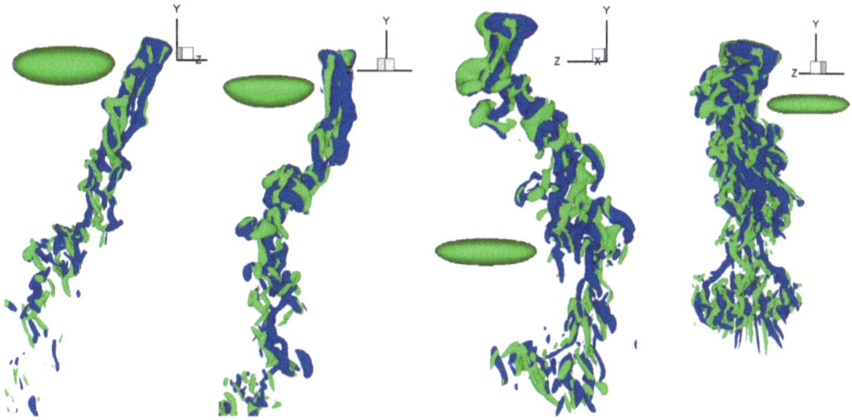

(a) $Eo = 0.74$ (b) $Eo = 1.21$ (c) $Eo = 2.2$ (d) $Eo = 4.9$

Fig. 5.4 The streamwise vortex structures behind the bubble. The parameters are as the same in Fig. 5.3

respectively. Obviously, the double-threaded wakes are observed to twist with one another, in accordance with some available studies [3, 6, 8] which conclude that the unstable wakes are mostly possible to cause the path instability. From Fig. 5.4a, c, typical interconnected hairpin-vortex structures are formed in the wakes of the bubbles. The similar vortex structures are also observed by Lunde and Peskin [11], who find that the wakes are consisted of two attached vortex filaments when the bubble rises in a spiral manner, and the hairpin-like vortex structures are generated simultaneously. This observation is then confirmed by Brücker [3], who thinks the zigzag motion of the bubble is caused by the alternative shedding of oppositely-oriented hairpin-like vortices. However, in Fig. 5.4d corresponding to the largest bubble, such vortex structures are not so identical, and this may be ascribed to the chaos motion of the rising bubble. Meanwhile, the bubble shape is observed to be more oblate as Eo increases, and according to the numerical predictions by Mougin and Magnaudet [16] and Shew and Pinton [20], larger aspect ratio produces more vorticities around the bubble interface, and correspondingly, the rising path of the bubble must be more unstable.

5.3 The Bubble Motion Under the Influence of a Vertical MF

In this section, we will investigate the influence of the vertical MF on the single Ar bubble motion in liquid GaInSn after we are aware of the relations between the bubble shapes, the bubble wakes and the bubble rising paths.

5.3.1 The Influence on the Rising Velocity

When the vertical MF is applied, another dimensionless parameter of the drag force coefficient in the form of $C_D = \frac{4}{3}\frac{\Delta\rho d_e g}{u_T^2 \rho_1}$ is introduced. Figure 5.5 shows the variation of C_D versus the interaction number N, while the experimental results [25] are presented for comparison.

In the figure, a striking difference is observed between the numerical and the experimental results, and this discrepancy is attributed to three aspects: (*a*) The liquid GaInSn used in the experiments is not pure because of the mixed oxide, which probably change the bubble motion behaviors and the influence needs to be further investigated. Regarding this issue, numerous experiments conducted in the contaminated water and the pure water already confirmed this [6, 23]. (*b*) The initial bubble shapes are not in fully agreement between the experiments and the numerical simulations, because the bubble released in the experiment may not be a spherical shape and this also affects its dynamic behaviors, as indicated by Wu et al. [24]. (*c*) The rising velocity measured in the experiments focuses on the lower part of the bubble, while we calculate the rising velocity by the motion of the bubble centroid. Nevertheless, the terminal velocity of the rising bubble when exposed to the vertical MF is, of course, our concern in the present work, however we also want to explore both the influence of the field and the mechanism involved.

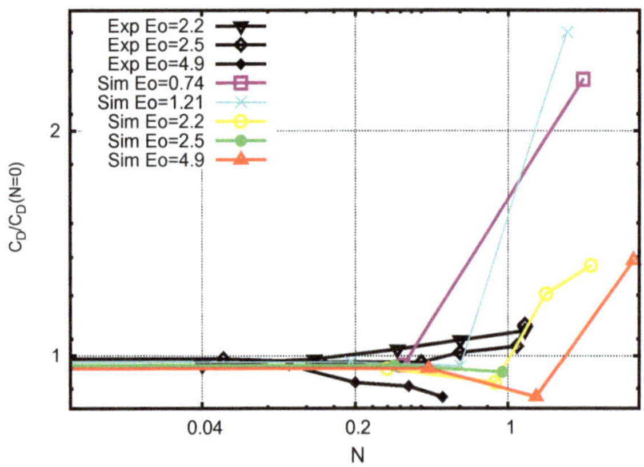

Fig. 5.5 Comparisons of C_D versus N between the numerical solutions and the experimental results [25] when the Ar bubble rises in presence of different vertical MFs

As concluded by Zhang et al. [25], small bubbles ($Eo < 2.5$) will be decelerated (larger C_D) by the vertical MF, while larger bubbles will gain speed (small C_D) by contraries. Before that, Mori [15] observed a similar phenomenon by investigating the N_2 bubble rising in mercury under a horizontal MF, he also found that the bubble would rise in a more rectilinear trajectory when the MF was strengthen, and correspondingly, the terminal velocity was increased. He then ascribed the increased velocity to the more straight rising path with the help of the horizontal MF.

Similarly, according to our numerical results, we find that the magnetic influence on the terminal rising velocities of the bubble is always non-monotonic regardless of the bubble sizes. The terminal velocity of the bubble will firstly increase and then decrease by strengthening the vertical MF gradually. However, for small bubbles, the critical value of N is also smaller and for larger bubbles, the critical N is increased accordingly. A detailed depiction of this is shown in Fig. 5.6, in which the rising velocity versus time are presented under different interaction parameters N, and the bubble sizes are $Eo = 0.74$, $Eo = 1.21$, $Eo = 2.2$ and $Eo = 4.9$, respectively.

From these figures, it is more clear that a smaller bubble will reach its maximum terminal velocity at smaller N. Alternatively, instead of using N, a new Eötvös number is defined based on the MF as $Eo_m = N \cdot We = \sigma_e u_T B^2 d/\sigma$ to establish

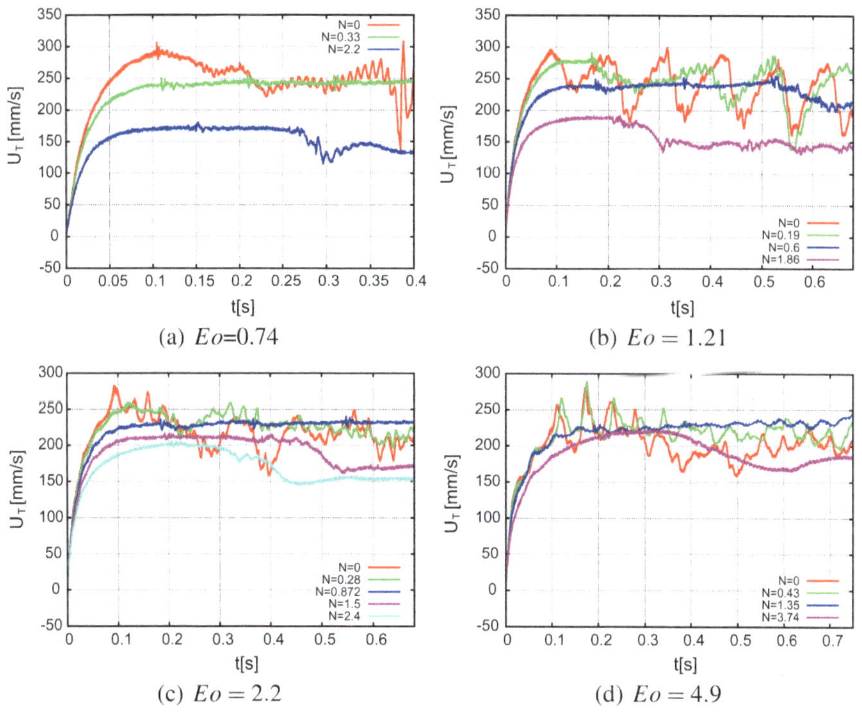

(a) Eo=0.74 (b) $Eo = 1.21$

(c) $Eo = 2.2$ (d) $Eo = 4.9$

Fig. 5.6 Time series of the rising velocity of different sized bubbles, the strengths of the vertical MFs is varied to investigate their influences

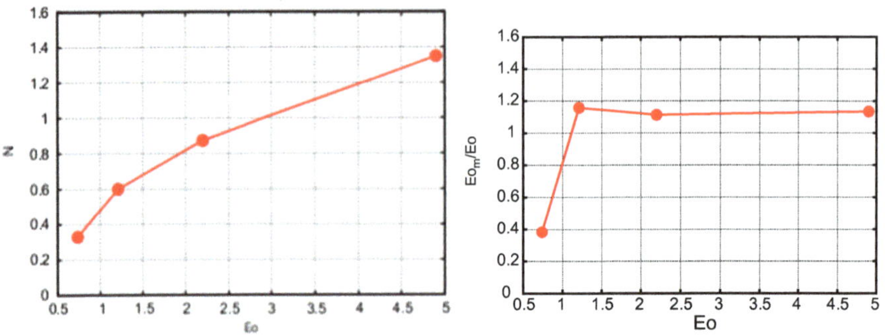

(a) $Eo \sim N$ (b) $Eo \sim Eo_m/Eo$

Fig. 5.7 Critical values of **a** Eo_m and **b** Eo_m/Eo when the peak terminal velocity appears by strengthening the vertical MF

a ratio between the gravitational and the magnetic forces. As shown in Fig. 5.7a, the bubble of $Eo = 0.74$ reaches its peak value at $Eo_m = 0.615 (N = 0.33)$, while $Eo_m = 1.41 (N = 0.6)$ for $Eo = 1.21$, $Eo_m = 2.61 (N = 0.872)$ for $Eo = 2.2$ and $Eo_m = 6.29 (N = 1.35)$ for $Eo = 4.9$. This shows that smaller bubble is more sensitive to the MF, indicating that its motion is not always suppressed by the vertical MF. Besides, because the Lorentz force acting on the bubble tends to balance the gravitational force, as a consequence, we try to define a new dimensionless parameter in the form of $Eo_m/Eo = \sigma_e u_T B^2 d/\Delta\rho g$ to represent the ratio between the two forces. Figure 5.7b gives the values of Eo_m/Eo when the peak terminal velocity appears by strengthening the vertical MF. As shown in the figure, a critical value of $Eo_m/Eo \approx 1.14$ is observed when the bubble size is larger than $Eo = 0.74$.

Meanwhile, as Fig. 5.6 shows, the terminal rising velocities are smoother under stronger MF, indicating that the oscillations of the rising velocity are suppressed significantly, this is also in good agreement with the experimental results.

5.3.2 *The Influence on the Rising Path*

As shown above, different vertical MF have contrary influence on the bubble motion: to promote it and to block it. According to the aforementioned numerical results in Sect. 5.3.1, the promotion effect dominates under weak MF, while the blocking effect is more predominant when the MF is stronger.

To confirm Mori's conjecture [15] that the variations of the bubble terminal velocity under different MFs are depended on the path transition, the bubble rising paths are investigated first. This means that by strengthening the vertical MF, the bubble rising trajectory is more rectilinear until it is fully straight, and before that, the bubble terminal velocity is increased with stronger MF. However, after the bubble rising path

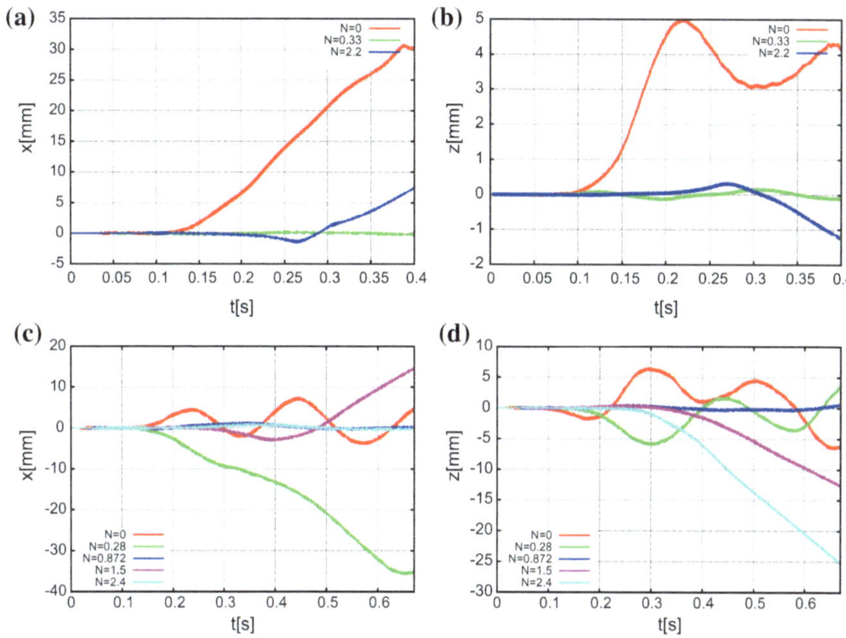

Fig. 5.8 Time histories of the bubble centroid in the x-(left column) and z-(right column) directions when various vertical MF are imposed. The bubble sizes are $Eo = 0.74$ (top) and $Eo = 2.2$ (bottom), respectively

is fully rectilinear, stronger vertical MF will further suppress the bubble motion. It can not be observed in the experiments because the liquid metal is opaque, however, it could be verified by performing numerical simulations.

This hypothesis is validated in Fig. 5.8, where the lateral drifts of the rising bubbles are shown in cases of $Eo = 0.74$ and $Eo = 2.2$. Regarding $Eo = 0.74$, it is clear that the bubble travels in an oblique trajectory (it only oscillates in the z-direction) without the applied MF, and a completely rectilinear path gradually forms till $N = 0.33$. In this period, the terminal rising velocity is observed to increase with stronger MF, as shown in Fig. 5.6a. After that, with a stronger MF from $N = 0.33$ to $N = 2.2$, its terminal rising velocity is decreased, as depicted in Fig. 5.6a.

More detailed investigations are given in case of $Eo = 2.2$, as presented in Fig. 5.8c, d. From the figures, the bubble travels in a spiral trajectory without the vertical MF (because it oscillates in both horizontal directions), then it travels in a zigzag path at $N = 0.28$, and the rising path finally becomes rectilinear at $N = 0.872$. Correspondingly, a gradual increase in the terminal velocity is observed in Fig. 5.6c during this stage, and simultaneously, the fluctuations in terminal velocity are also suppressed. After that, greater magnetic intensities of $N = 1.5$ and $N = 2.4$ will lead to reduced terminal velocities. Again, it indicates that after the bubble rises in a fully straight path under the influence of the vertical MF, stronger MF will block the

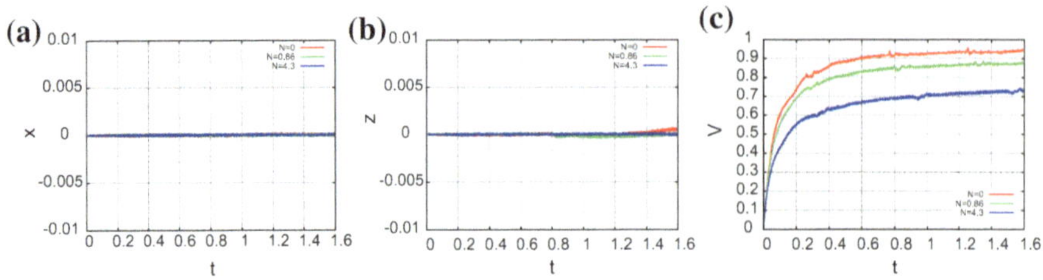

Fig. 5.9 In case of rectilinearly rising bubble without MF, the figure respectively shows: **a** the drift of the bubble centroid along the x-direction; **b** the drift of the bubble centroid along the z-direction; and **c** time histories of the rising velocity. It is observed that the bubble is always blocked by the vertical MF

bubble rising by contraries. Nevertheless, it should be noted that the bubble deflects from its vertically straight trajectory with larger N, and the originally stable rise velocity will drop slightly when this path instability occurs, as shown in Fig. 5.6c when the magnetic strength is larger than $N = 1.5$. Such "second instability" under the influence of the vertical MF also appears in case of $Eo = 0.74$, and it will be discussed in details in the next section.

In addition, to identify that the reverse influences of the vertical MF on the bubble motion are depended on whether the bubble rises in a rectilinear path or not, another group of numerical simulations are conducted, focusing on the magnetic influence on the rectilinearly rising bubble. By ensuring $Re_T = 100$ and $We_T = 4.5$ in the simulations, the numerical results are presented in Fig. 5.9, where the first two figures show the lateral drifts of the bubble centroid, while the last picture is the histories of the rising velocity under different N. It is observed that under such circumstance, the bubble motion is always suppressed by the vertical MF.

5.3.3 The Drag Force on the Bubble Motion

Now we are aware of why a weak magnetic field will accelerate the bubble motion, that is because the MF tends to straighten the rising path of the bubble, and the mechanisms will be discussed in the next section by investigating the evolution of the vortex structures behind the bubble. Herein, we would like to discuss how the drag effect comes from when the vertical MF is imposed, and the rectilinearly rising case of $Re = 100$ is given for instance. We should keep in mind that by $\boldsymbol{F} = \boldsymbol{J} \times \boldsymbol{B}$, the vertically directed MF does not produce a vertical drag force to block the bubble motion, therefore, the drag force must come from elsewhere.

Firstly, the vectors of the current density and the Lorentz force in case of $N = 4.3$ are plotted in Fig. 5.10, where the MF is vertically directed. It is obvious that the current densities flow in the counterclockwise direction at the upper part of the bubble, while they flow in the clockwise direction at the lower part. Consequently, the

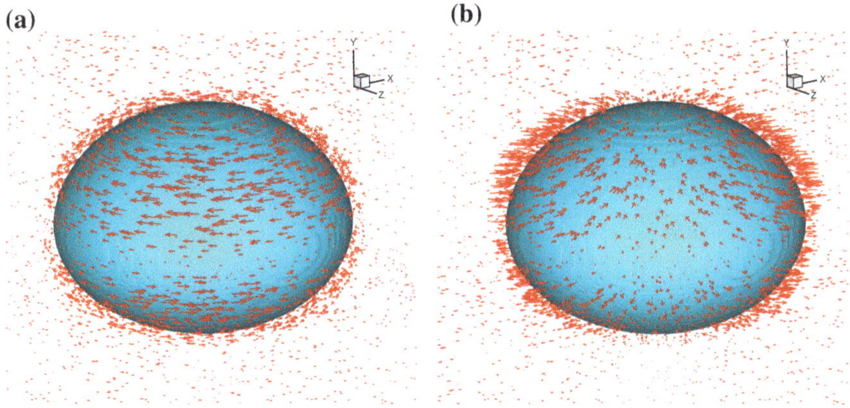

Fig. 5.10 The distribution of **a** the current density vectors and **b** Lorentz force around the bubble, the terminal rising velocity results in $Re = 100$ and $N = 4.3$

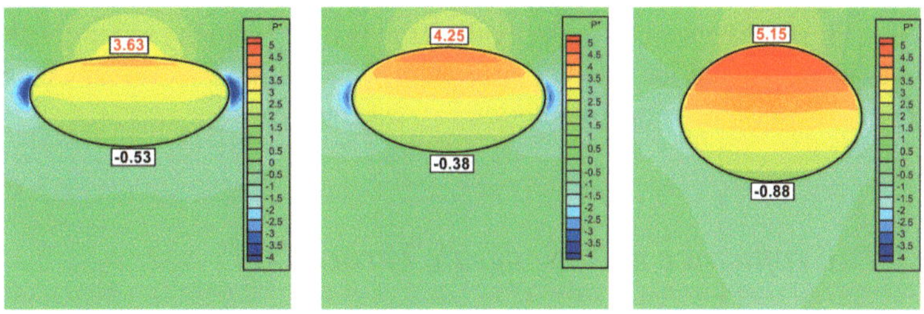

(a) $N = 0$ (b) $N = 0.86$ (c) $N = 4.3$

Fig. 5.11 The pressure distributions around the bubble under different vertical MFs, and the labeled numbers indicate the corresponding pressure at the top and the bottom of bubble. The cases correspond to that in Fig. 5.9

induced Lorentz force has opposite directions too, respectively of acting towards and outwards of the bubble centroid at the upper and lower parts of the bubble, as shown in Fig. 5.10b. Therefore, such distributions of the Lorentz force result in a higher pressure at the upper part and a lower pressure at the lower part correspondingly.

Under the action of the Lorentz force, the pressure difference between the top and the bottom of the bubble is expected to increase with stronger MF, which induces larger drag force. To confirm this, Fig. 5.11 shows the pressure distributions at the top and the bottom of the bubble when the vertical MF is varied. It should be noted that in the figures, the hydrodynamic pressure are taken away, being $p^* = p - \rho_l g H$, where ρ_l denotes the liquid density and H is the distance from the top wall. The labels in the figures show p^* at the top and the bottom of the bubble that a larger pressure difference is observed when N is larger, and thus a larger drag force is generated to block the bubble motion.

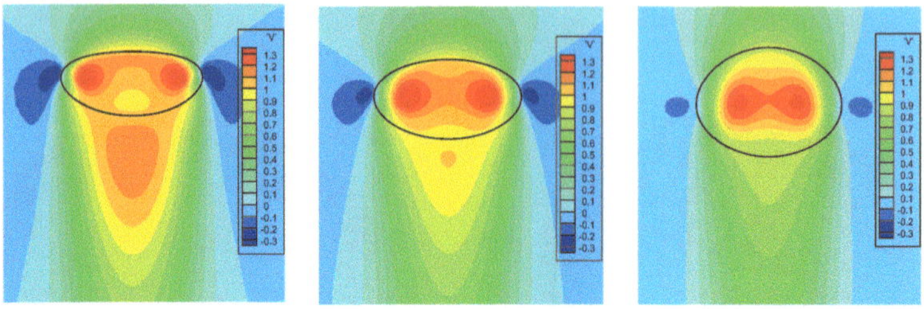

(a) $N = 0$ (b) $N = 0.86$ (c) $N = 4.3$

Fig. 5.12 The contour maps of the rising velocity in vicinity of the bubble under different vertical MFs. The cases correspond to those in Fig. 5.9

In addition, as Fig. 5.11 shows, the bubble becomes more spherical and it changes to a bullet-like shape under stronger MF. This is in good agreement with some available numerical results, given by Takatani et al. [22] and Shibasaki et al. [21].

Finally, Fig. 5.12 presents the distributions of the rising velocity around the bubble. Under stronger MF, the jet velocity of the liquid below the bubble will reduce and therefore the effect on the bubble motion is weakened.

5.3.4 *The Influence of a Moderate Vertical MF*

As described in Sect. 5.1, it is widely known that the path instability of a single bubble motion is triggered by the wake instability, and numerous experimental and numerical studies performed in water have confirmed this [3, 6, 23]. However, few direct numerical simulations are available to provide more detailed information about the evolution of the vortex structures during the transition of the path instability. Also, when the vertical MF is applied on the rising bubble, the unstable trajectory becomes more rectilinear, as presented in the Sect. 5.3.2, however the in-depth physical mechanisms for this transition still need to be identified. Being inspired by the importance of the wake instability, we attempt to present the evolution of the vortex structures behind the bubble when the vertical MF is varied. Particularly, the bubbles of fixed $Eo = 2.2$ are analyzed in the following study, because we believe the magnetic influence on the vortex structures must be similar.

To begin with, the evolution of the vortex structure behind a bubble without MF is presented that two time instants are selected: at $t = 0.1$ s when the path is still rectilinear and at $t = 0.4$ s when the spiral trajectory forms, as shown in Fig. 5.13. The iso-contours indicate the streamwise vorticities to be $\omega_y = \pm 200\,\mathrm{s}^{-1}$. In Fig. 5.13a, the vortex structures are stable when the bubble rises in a rectilinear path, however, as the spiral motion is formed in Fig. 5.13b, a more complex vortex structure is observed that the double-threaded vortices tend to twist with one another. In addition, the vorticities are observed to be shed in the tail of the wake, indicating that the vortex

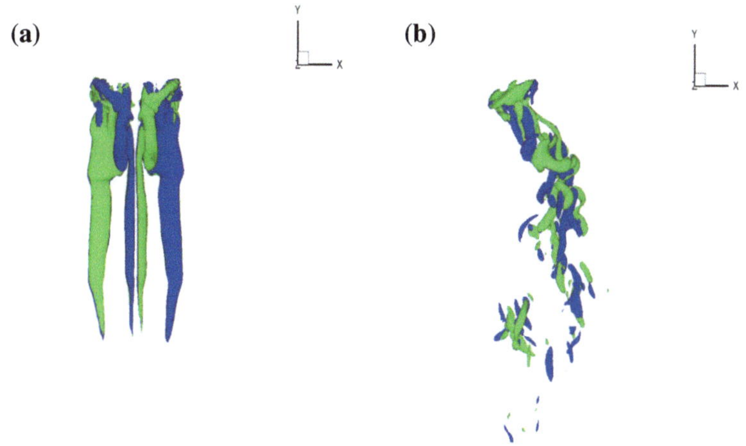

Fig. 5.13 The vortex structures behind the rising bubble without MF, the bubble rises within a **a** rectilinear and **b** spiral motion, respectively at $t = 0.1$ s and $t = 0.4$ s. The iso-contours correspond to $\omega_y = \pm 200\,\mathrm{s}^{-1}$, and the bubble size is $Eo = 2.2$

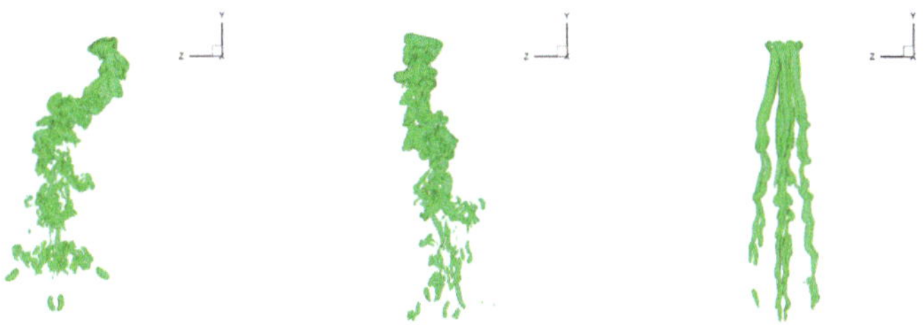

Fig. 5.14 The vortex structures behind the bubble under different MFs, respectively of $N = 0$, $N = 0.28$, and $N = 0.872$ and they are all at the time instant of $t = 0.4$ s. The bubble size is fixed at $Eo = 2.2$

structures are not stable any longer. As a consequence, the numerical results confirms that the path instability and the wake instability are closely related.

After that, the vortex structures are presented in Fig. 5.14 under different vertical MFs, respectively of $N = 0$, $N = 0.28$, and $N = 0.872$. As shown in the figures, the vortex strengths are weakened by stronger MF in succession while the vortex filaments are apt to be parallel with the MF gradually. Accompanied by such vortex evolutions, the bubble also tends to rise in a more rectilinear way, corresponding to Fig. 5.8c, d. Consequently, it is proved that by modifying the wake structures behind the bubble, its rising trajectory is also changed. In addition, this effect on the bubble wake is in good agreement with the theoretical predictions by Davidson [5], who declared that the flow field will be more homogeneous in direction parallel to the field.

5.3.5 The Influence of a Strong Vertical MF

In this section, we provide a more comprehensive discussion about the "second path instability" under stronger MF in case of $N = 1.5$ and $N = 2.4$ in case of $Eo = 2.2$, as shown in Fig. 5.8c, d. Obviously, this kind of path instability is different from that under $N = 0$ and $N = 0.28$, because the "second path instability" simply shows an oblique trajectory without path oscillations. In particular, the discussions in case of $N = 2.4$ are presented in details.

Similarly, two time instants of $t = 0.2$ s and $t = 0.3$ s are selected, typically before and after the "second path instability" happens. The distribution of the current density streamlines in vicinity of the bubble are presented in Fig. 5.15, where the front and the back views are displayed, respectively. It is apparent that when the bubble travels in a rectilinear path at $t = 0.2$ s , the current densities are distributed symmetrically at both $\pm z$ ends of the bubble. However, this balance is broken at $t = 0.3$ s that an asymmetric distribution of the current densities is observed, therefore unequal Lorentz forces are then produced at $\pm z$ ends of the bubble to push the bubble to deviate from the rectilinear path.

Moreover, the vortex structures at $t = 0.23$ s and $t = 0.3$ s are presented, as shown in Fig. 5.16 that no twisted vortex filaments are observed, even when the bubble rises in a oblique line at $t = 0.3$ s. This is quite different from the cases of $N = 0$ and $N = 0.28$, by which the path instability is related to the vortex instability. Therefore, the "second path instability" under a strong vertical MF is not triggered by the wake

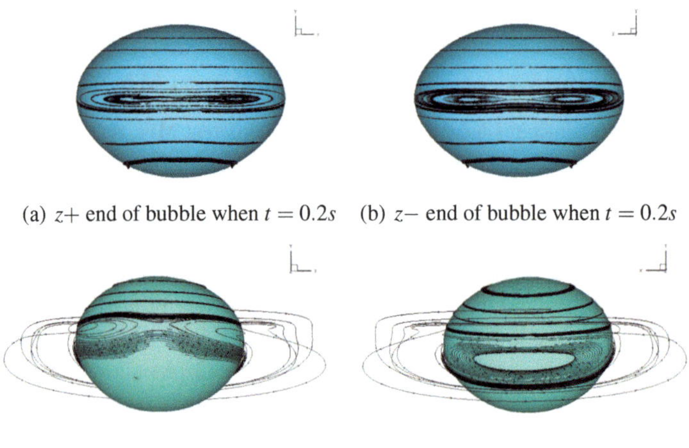

(a) $z+$ end of bubble when $t = 0.2s$ (b) $z-$ end of bubble when $t = 0.2s$

(c) $z+$ end of bubble when $t = 0.3s$ (d) $z-$ end of bubble when $t = 0.3s$

Fig. 5.15 Current density distributions around the bubble at different stages, **a**, **b** before and **c**, **d** after the "second path instability" happens. The corresponding parameters are correspondent to $Eo = 2.2$ and $N = 2.4$. It is observed that the symmetrical distribution of the current density is broken when the "second path instability" happens

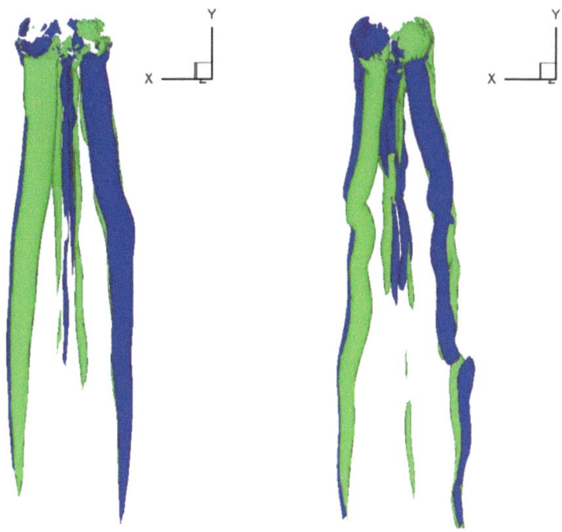

(a) $t = 0.2\,s$ (b) $t = 0.3s$

Fig. 5.16 Vortex structures behind the bubble at different stages, **a** before and **b** after the "second path instability" happens. The parameters in this case are identical to those in Fig. 5.15

instability. Furthermore, after a more detailed inspection on the vortex structures at the two time instants, the vortex thread is observed to break up when the bubble begins to deviate from the rectilinear path at $t = 0.23$ s, while the asymmetrical vortex structure is more obvious at $t = 0.3$ s that the vortex line at the z-end fully vanishes. This must be attributed to the asymmetric distribution of the Lorentz force.

We attribute this asymmetric distribution of the Lorentz force to the instability in high Re flows. It is known that this instability occurs most probably at the bubble waist [20], in which region a small disturbance may cause the asymmetrical distribution of the current density, as shown in Fig. 5.15. In addition, by comparing the results of $N = 1.5$ and $N = 2.4$, a stronger vertical MF is found to induce the "secondary path instability" more easily. After that, we increase N to a much larger value, i.e. $N = 65$, by which the flow instability is greatly suppressed under the MHD effect. Figure 5.17 shows the lateral drifts of the bubble centroid in this case. Obviously, it shows that the bubble will never deviate from the rectilinear path in case of $N = 65$, indicating that the "second path instability" is disappeared. This is probably attributed to the more stable flow under strengthened MF, because the Reynolds number is reduced to $Re = 1400$. In summary, we come to a short conclusion on this topic: *(a)* a smaller vertical MF will cause the bubble to rise in a more rectilinear manner; *(b)* beyond a critical value of N, the rising bubble will show a "secondary path instability" which is attributed to the flow instability, which will in turn induce an asymmetrical distribution of the Lorentz force; *(c)* however, a larger N will stabilize the bubble motion again, because the flow instability is greatly suppressed by the MF, and the bubble trajectory then returns to a rectilinear line.

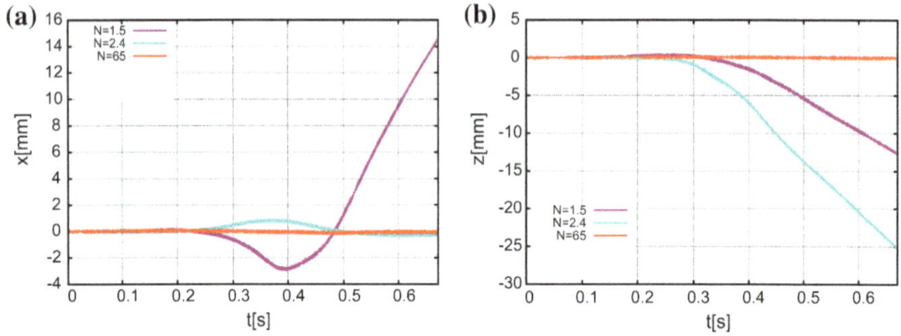

Fig. 5.17 Time histories of the bubble centroid along the **a** x- and **b** z-directions under much stronger vertical MF of $N = 65$, while the results under $N = 1.5$ and $N = 2.4$ are also attached for comparison. It is observed in case of $N = 65$, the bubble rises in a rectilinear path because of more stable flows

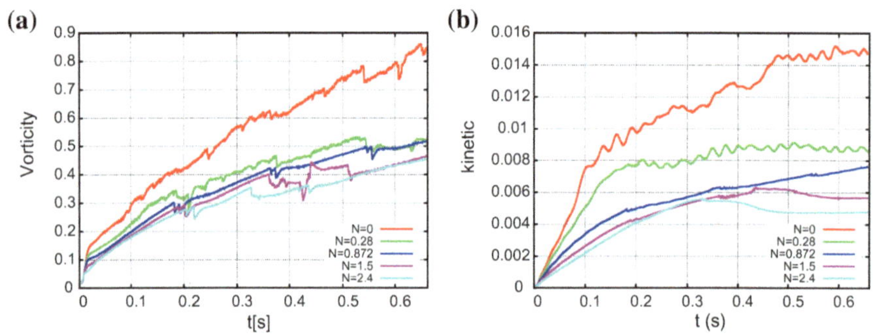

Fig. 5.18 Time histories of total summation of **a** the vorticity and **b** the kinetic energies over the whole computational domain, the bubble size is $Eo = 2.2$. It is observed that both quantities are suppressed by the vertical MF

At last, the histories of the total summation of the vorticities over the whole computational domain are presented in Fig. 5.18a. A monotonic decrease tendency is observed with stronger vertical MF. Similarly, by tracking the development of the kinetic energies over the computational domain, as shown in Fig. 5.18b, we observe a similar trend which indicates that the vertical MF acts to dissipates the energy by Joule dissipating.

5.3.6 *The Influence on the Bubble Shape*

Regarding the influences of the vertical MF on the bubble, another characteristics that should be taken into consideration is the aspect ratio of the shape. As Zenit and Magnaudet [26] argue in their experimental study, it is the bubble shapes rather

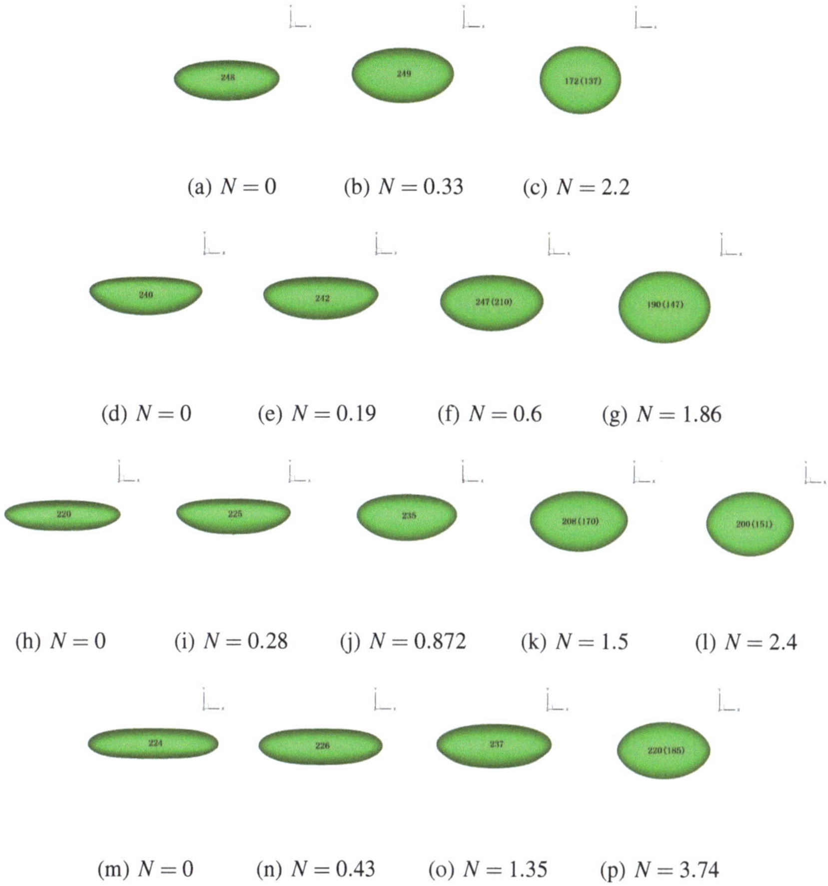

Fig. 5.19 Variations of the bubble shape and terminal velocity under the influence of vertical MF, respectively of **a–c** $Eo = 0.74$, **d–g** $Eo = 1.21$, **h–l** $Eo = 2.2$ and **m–p** $Eo = 4.9$

than Re to trigger the vortex instability because the surface vorticity would exceed a certain threshold in case of larger χ. Mougin and Magnaudet [16] also confirm that larger aspect ratio of the rising bubble will lead to more vorticity accumulation, which leads to path instability directly. The relationship between the production of the vorticity and the maximum curvature has been discussed in another two papers [4, 26] and consequently, the bubble shapes as functions of Eo and N have been plotted in Fig. 5.19 before the path instability happens. Meanwhile, the terminal velocity corresponding to each case is also labeled in each sub-figure with unit of mm/s, while the brackets differentiate the values before and after the "second path instability" takes place.

Without the vertical MF, the bubble becomes more flat as the bubble size increases, and hence larger χ is observed. Therefore, more vorticities are produced in vicinity of

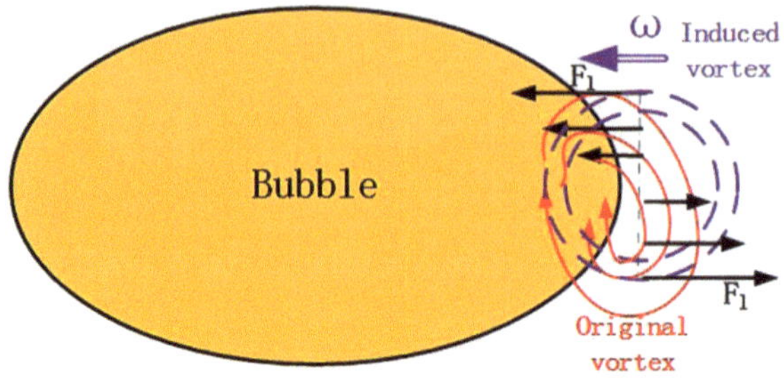

Fig. 5.20 The influence of the vertical MF on the pressure distributions in vicinity of the bubble

Table 5.3 Maximum vorticity at the bubble surface under different vertical MF, the bubble size is $Eo = 2.2$

N	0	0.28	0.872	1.5	2.4
ω_{max}	5000	4000	2000	1400	1000

a larger bubble because the curvature is also larger. This is also validated in Fig. 5.4, and it is consistent with either the available experimental or numerical results.

In addition, as the vertical MF is intensified, a smaller aspect ratio is observed in Fig. 5.19, regardless of the bubble size. This change trend of the shape deformation of the bubble will inevitably produce fewer vorticities at its interface because of smaller curvature. To investigate why the bubble shape evolves under the influence of the vertical MF, the distribution of the Lorentz force in vicinity of the bubble is presented in Fig. 5.20. As shown in the figure, the Lorentz forces at the upper and the lower part of the bubble act in opposite directions, and hence a torque is induced to weaken the original vortex induced by pressure difference, and consequently the pressure at the bubble twist is increased. On the other hand, the surface tension, to be balanced by the pressure difference between the interior and the exterior of the bubble as $\sigma\kappa = p_{in} - p_{out}$, will decrease when the pressure is elevated and hence a smaller curvature there is resulted.

To validate the above conjecture, the maximum vorticity at the bubble interface is plot versus different N, as shown in Table 5.3 and a decreased tendency is observed.

Another issue needs to be discussed during the bubble motion is the shape oscillations, which can be characterized by the length of three axes d_1, d_2 and d_3, as schematically sketched in Fig. 5.21. The evolution of d_1, d_2 and d_3 are shown in Fig. 5.22 under the influence of vertical MFs, again in case of $Eo = 2.2$. In Fig. 5.22a without MF, d_1 and d_2 elongates as the bubble starts to rise, while d_3 is shortened. After $t^* = 0.1$ s, a relatively stable aspect ratio of $\chi = 3.6$ is reached, however the small fluctuations indicate the periodic shape oscillations. In case of $N = 0.28$, the terminal aspect ratio is decreased to $\chi = 3.2$, and also the fluctuations are also suppressed by the vertical MF. Subsequently in case of $N = 0.872$, no fluctuations are observed on the length of the axes after the aspect ratio grows to $\chi = 2.2$, and the shape oscillations are fully eliminated.

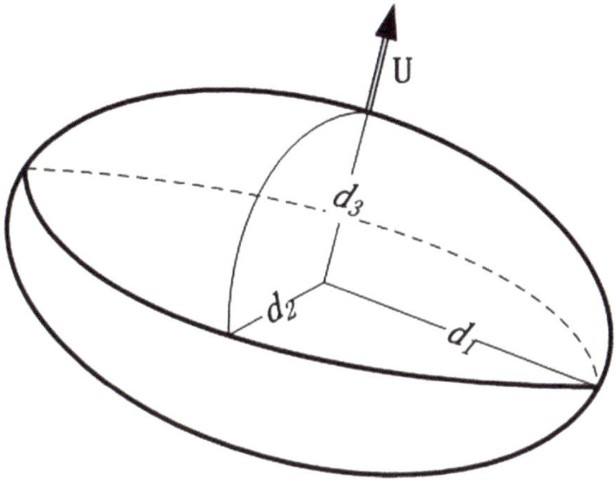

Fig. 5.21 Schematic diagram defining the axes of the rising bubble

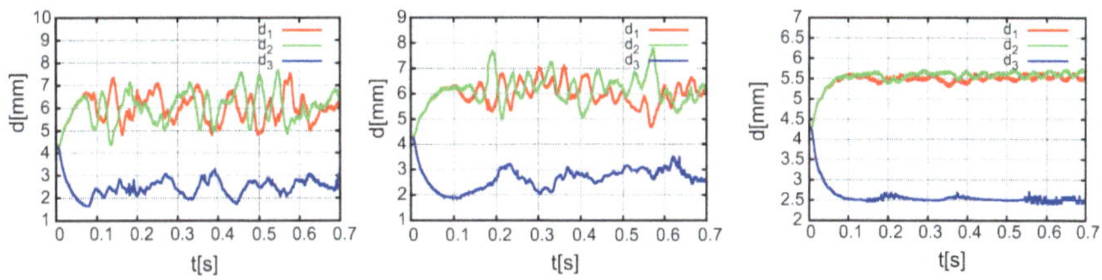

(a) $N = 0$ (b) $N = 0.28$ (c) $N = 0.872$

Fig. 5.22 The shape oscillations of the rising bubble under the influence of vertical MFs, the bubble size is $Eo = 2.2$

5.4 Summary

In this chapter, the numerical methodologies developed in Chap. 3 are used to simulate a single argon bubble rising in liquid GaInSn, and the influence of the vertical MF is investigated. To be most visibly, the numerical results show that the vertical MF has a non-monotonic effect on the terminal velocity of the bubble, indicating that a moderate MF will promote the bubble rising, while a strong MF suppresses its motion. Moreover, the critical strength of the vertical MF is depended on the bubble size that a smaller bubble also corresponds to a weaker critical MF. This discovery is slightly different from the experimental results by Zhang et al. [25], who find the vertical MF always acts to suppress the motion of the small bubble. Nevertheless, quantitative comparison between our numerical solutions and the experimental results show significant differences in the terminal velocities under varied vertical MFs, this is probably attributed to the impurities of mixed oxides in the experimental liquid GInSn.

To investigate why the vertical MF tends to promote the bubble motion in a certain range, the variations of the rising path of the bubble are studied at first. It is observed that under a moderate MF, the rising trajectory of the bubble will be corrected from a spiral manner into a more rectilinear way, and correspondingly, the terminal velocity of the bubble increases. After that, a stronger MF would reduce the terminal velocity by contraries, and the MF begins to prevent the bubble motion. This discovery is consistent with the conjecture put forward by Mori, who conducted the similar experiments by using mercury. Consequently, the vertical MF has opposite influence on the bubble motion: the positive effect is dominant when N is moderate, while the negative effect, which is resulted from the enlarged pressure difference between the top and the bottom of the bubble due to Lorentz force, is more obvious for larger N.

We also observe the second path instability to occur and vanish by intensifying the vertical MF continuously, and a detailed investigation indicates such a path instability is caused by the asymmetric distribution of the Lorentz force around the bubble because of the flow instability.

Motivated by the available numerical and experimental results that an unstable bubble path is triggered by the wake instability, we then investigate the variations of the vortex structures under the influence of vertical MF. It is found that the MF modifies the vortex structures behind the bubble wake remarkably, such that the twisted vortex threads become more regular and tend to be parallel with the MF. It is shown that the reorganization of the vortex structure under the influence of vertical MF is related to the more rectilinear rising path of the bubble. Meanwhile, both the vorticity and the kinetic energy over the computational domain are suppressed by the vertical MF, indicating to the influence of the Joule dissipation. In addition, the "second path instability" is also found to be attributed to the asymmetrical vortex structure, which induces an unbalanced Lorentz force to deviate the bubble from the rectilinear path.

Finally, the shape deformations under the influence of the vertical MF are discussed, the results show that the aspect ratio of the bubble tends to be smaller as N increases, and such variation trend is believed to produce fewer vorticities at the bubble interface. Moreover, the shape oscillation of the bubble is also suppressed by the vertical MF.

References

1. Bai H, Thomas BG (2001) Bubble formation during horizontal gas injection into downward-flowing liquid. Metall Mater Trans B 32(6):1143–1159
2. Benjamin TB (1987) Hamiltonian theory for motions of bubbles in an infinite liquid. J Fluid Mech 181:349–379
3. Brücker C (1999) Structure and dynamics of the wake of bubbles and its relevance for bubble interaction. Phys Fluids 11(7):1781–1796
4. Cano-Lozano JC, Bohorquez P, Martnez-Bazn C (2013) Wake instability of a fixed axisymmetric bubble of realistic shape. Int J Multiph Flow 51:11–21

5. Davidson PA (1995) Magnetic damping of jets and vortices. J Fluid Mech 299:153–186
6. De Vries AWG, Biesheuvel A, Van Wijngaarden L (2002) Notes on the path and wake of a gas bubble rising in pure water. Int J Multiph Flow 28(11):1823–1835
7. Ellingsen K, Risso F (2001) On the rise of an ellipsoidal bubble in water: oscillatory paths and liquid-induced velocity. J Fluid Mech 440:235–268
8. Gaudlitz D, Adams NA (2009) Numerical investigation of rising bubble wake and shape variations. Phys Fluids 21(12):122102
9. Gaudlitz D, Adams NA (2010) The influence of magnetic fields on the rise of gas bubbles in electrically conductive liquids. Direct and large-eddy simulation VII, pp 465–471
10. Hartunian RA, Sears WR (1957) On the instability of small gas bubbles moving uniformly in various liquids. J Fluid Mech 3(1):27–47
11. Lunde K, Perkins RJ (1997) Observations on wakes behind spheroidal bubbles and particles. In: ASME Fluids Engineering Division Summer Meeting, Vancouver, Canada, p 141
12. Magnaudet J, Eames I (2000) The motion of high-Reynolds-number bubbles in inhomogeneous flows. Annu Rev Fluid Mech 32(1):659–708
13. Mendelson HD (1967) The prediction of bubble terminal velocities from wave theory. AIChE J 13(2):250–253
14. Miao X, Lucas D, Ren Z et al (2013) Numerical modeling of bubble-driven liquid metal flows with external static magnetic field. Int J Multiph Flow 48:32–45
15. Mori Y, Hijikata K, Kuriyama I (1977) Experimental study of bubble motion in mercury with and without a magnetic field. J Heat Transf 99(3):404–410
16. Mougin G, Magnaudet J (2001) Path instability of a rising bubble. Phys Rev Lett 88(1):014502
17. Ni MJ, Munipalli R, Morley NB et al (2007) A current density conservative scheme for incompressible MHD flows at a low magnetic Reynolds number. Part I: on a rectangular collocated grid system. J Comput Phys 227(1):174–204
18. Ni MJ, Munipalli R, Huang P et al (2007) A current density conservative scheme for incompressible MHD flows at a low magnetic Reynolds number. Part II: on an arbitrary collocated mesh. J Comput Phys 227(1):205–228
19. Saffman PG (1956) On the rise of small air bubbles in water. J Fluid Mech 1(3):249–275
20. Shew WL, Pinton JF (2006) Dynamical model of bubble path instability. Phys Rev Lett 97(14):144508
21. Shibasaki Y, Ueno K, Tagawa T (2010) Computation of a rising bubble in an enclosure filled with liquid metal under vertical magnetic fields. ISIJ Int 50(3):363–370
22. Takatani K (2007) Mathematical modeling of incompressible MHD flows with free surface. ISIJ Int 47(4):545–551
23. Veldhuis C, Biesheuvel A, Van Wijngaarden L (2008) Shape oscillations on bubbles rising in clean and in tap water. Phys Fluids 20(4):040705
24. Wu M, Gharib M (2002) Experimental studies on the shape and path of small air bubbles rising in clean water. Phys Fluids 14(7):L49–L52
25. Zhang C, Eckert S, Gerbeth G (2005) Experimental study of single bubble motion in a liquid metal column exposed to a DC magnetic field. Int J Multiph Flow 31(7):824–842
26. Zenit R, Magnaudet J (2008) Path instability of rising spheroidal air bubbles: a shape-controlled process. Phys Fluids 20(6):061702

Chapter 6
The Argon Bubble Rising in the Liquid GaInSn Under the Influence of a Horizontal Magnetic Field

After a previous investigation of the rising motion of a gas bubble in the liquid GaInSn under the influence of a vertical MF, this chapter focuses on the case of a uniform horizontal MF. Similarly, in order to allow some benchmark, most of the parameters selected for this new investigation are the same as in an experiment recently performed in Dresden, Germany. The Ar bubble diameter is either 4.3 mm or 6.4 mm, and the liquid metal is still GaInSn, resulting in Reynolds numbers in the parameter space of $2000 < Re < 4000$, which is typically larger than that rising in water. The research is carried out in a similar manner, by investigating the influence of the horizontal MF on the rising velocity, the rising path, the vortex structures and the shape of the bubble, and finally, we give a brief comparison of the numerical results between the vertical and horizontal MFs.

6.1 The Bubble Motion Under the Influence of a Horizontal MF

6.1.1 *Suppression of the Axial Symmetry*

This anisotropy in the horizontal planes is the first remarkable effect observed by Miao et al. [4] and Zhang et al. [10] in their experiments. Let us therefore start with our numerical data on this particular aspect. A Cartesian frame of reference besides the upward vertical direction of Oy, Ox the direction of the horizontal MF and Oz the transverse direction perpendicular to the MF, is used throughout this chapter to distinguish the different horizontal directions. The selected Eötvös number is the same as in the Zhang experiment ($Eo = 2.2$). Three typical values of the interaction parameter are considered in order to exhibit `(i)` the fluid flow without any MF

J. Zhang, *The Developments and the Applications of the Numerical Algorithms in Simulating the Incompressible Magnetohydrodynamics with Complex Boundaries and Free Surfaces*, Springer Theses, https://doi.org/10.1007/978-981-10-6340-4_6

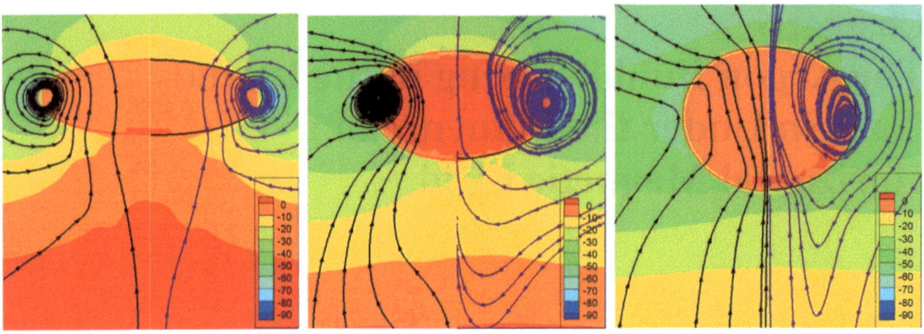

(a) $N = 0$ (b) $N = 3.6$ (c) $N = 24$

Fig. 6.1 Sketches of the velocity and pressure distributions without and with a horizontal MF: in the (Ox, Oy) plane, parallel to the magnetic field, on the left-hand side; in the (Oy, Oz) plane, perpendicular to the magnetic field, on the right hand side. The plotted lines are projections of the three-dimensional streamlines in the bubble vicinity. The color scale illustrates the pressure distribution

($N = 0$), (ii) a regime where inertia and Lorentz force have similar orders of magnitude ($N = 3.6$), and (iii) a regime where Lorentz force is predominant ($N = 24$). Figure 6.1 shows the projections of the streamlines on either the (Ox, Oy) plane containing the MF direction (left hand side half-picture), or the (Oz, Oy) plane perpendicular to the MF (right hand side half-picture).

The case $N = 0$ (Fig. 6.1a) presents the liquid metal flow at time such that the bubble trajectory is still rectilinear (as shown in the following, instabilities start at $t \approx 0.1$ s and the total rising duration is 0.7 s). As expected, it shows a perfect symmetry between the Ox and the Oz directions and some flattening of the bubble. It is clear that the liquid metal is driven up in the central region, both upstream and downstream the rising bubble, and that it moves downwards all around this region, thus generating a toroidal vortex centred on the equatorial bubble border. The descending flow is driven by the adverse pressure gradient, which is illustrated by the color scale where the numerical values are given in SI units.

For cases $N = 3.6$ and $N = 24$, similar projected streamlines are plotted at time $t = 0.16$ s when the bubble has reached its stable rising velocity. The descending liquid metal flow in the plane parallel to the MF is strongly reduced, and even tends to be suppressed when N gets very large. On the contrary, in the plane perpendicular to the MF, the toroidal vortex is enlarged and the recirculating flow rate is larger.

The key to understand this reorganization of the liquid metal flow in the presence of a strong horizontal MF is the competition between the Lorentz force, proportional to B^2 since the current density is proportional to B, and the pressure gradient. With a moderate MF, inertia also plays a significant role in this force balance, but it becomes smaller and smaller when N increases. Let us therefore consider the case $N = 3.6$, where the Lorentz force is large enough to allow disregarding inertia. The projection of the induced electric current lines on the bubble top surface is shown on Fig. 6.2a and the resulting Lorentz force distribution is shown on Fig. 6.2b. It can be noticed

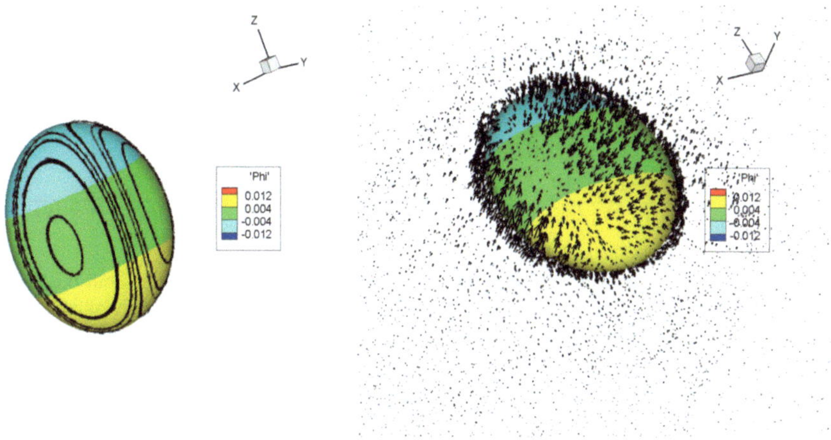

(a) Distributions of current density (b) Distributions of Lorentz force

Fig. 6.2 Computed electric current lines and Lorentz forces on the upper free surface of the bubble in the intermediate case $B = 3.6$. The color scale illustrates the electric potential distribution

that the current lines form closed loops, which confirms that the electric charge conservation is well satisfied at the whole bubble scale. The current density on the bubble surface is oriented in the direction opposite to Oz with a small value near the highest part, and it is oriented in the Oz direction near the equatorial circle, with larger values. This results from the equilibrium between the electromotive field $\boldsymbol{u} \times \boldsymbol{B}$ and the electric field $\boldsymbol{E} = -\nabla\varphi$, which is generated by the electric charges accumulated at the insulating bubble interface. The color scale allows getting an idea of the distribution of the electric potential. As a consequence, in the vicinity of the equatorial circle, the Lorentz force is large and oriented upwards in the direction of the MF (Ox), whereas it is small and oriented downwards in the perpendicular direction (Oz). This is the cause for the anisotropy observed on Fig. 6.1b, c.

The distribution of both the pressure gradient $\nabla p^* = \nabla(p + \rho_l g y)$ and the Lorentz force, which are competing with each other, is shown on Fig. 6.3. It appears that, when N becomes large enough, the vertical component of this pressure gradient is still predominant at large distance from the bubble. On the contrary, the vertical Lorentz force is predominant in the immediate bubble vicinity and on its diameter in the MF direction. In such a situation, the competition between those two forces generates the vortex sketched on Fig. 6.4. This vortex cannot be toroidal since the driving mechanism is only present in the vicinity of the plane parallel to the MF. On the contrary, in the plane perpendicular to the MF, both the Lorentz force and the pressure gradient are oriented downwards, as shown on Fig. 6.2b. It is worth emphasizing that this competition between Lorentz force and pressure gradient is quite different from the mechanism explaining the axially symmetric vortex present when $N = 0$, due to the competition between pressure gradient and inertia.

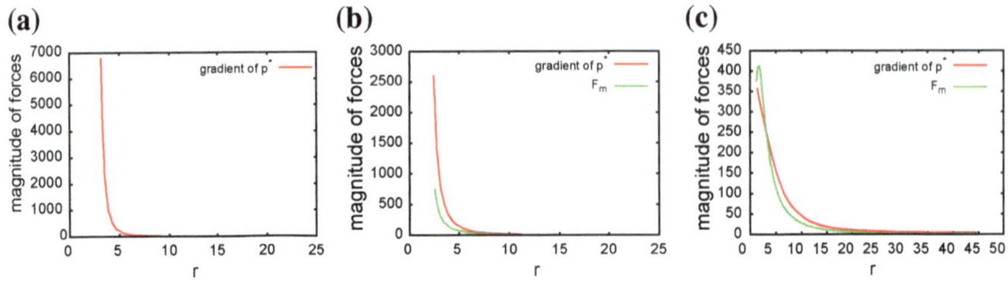

Fig. 6.3 Comparison between the distributions of the vertical components of both the Lorentz force and the pressure gradient in the mid-plane of the bubble as functions of the dimensionless distance from the bubble centre, along with the Ox-direction showing that Lorentz force becomes predominant in the immediate bubble vicinity when N is large enough. **a** $N = 0$, **b** $N = 3.6$, and **c** $N = 24$

Fig. 6.4 Schematic illustration of the competition between Lorentz force and pressure gradient, generating vortices at both ends of the bubble diameter oriented along the Ox−axis

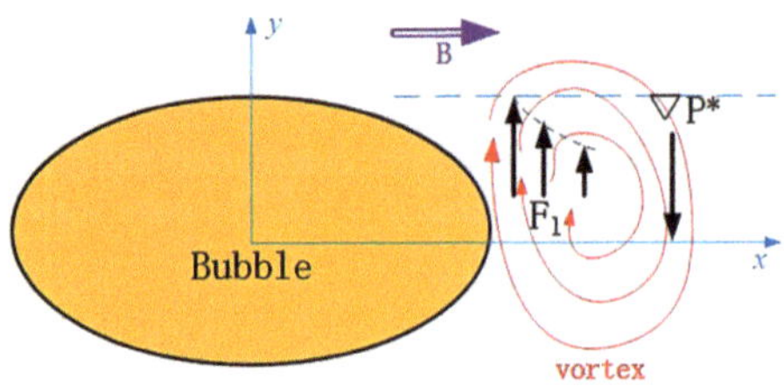

One might also remark that the flow organization around the rising bubble in the presence of a horizontal MF agrees fairly well with the Zhang et al. [11] ultrasonic measurements of the velocity profiles. It also fits well with the general ideas on the necessary development of some anisotropy in MHD flows submitted to a uniform MF, for instance presented in papers by Sommeria and Moreau [8] and Davidson [2]. In short, these theoretical investigations show that the Lorentz force tends to reduce the velocity gradient in the direction parallel to the MF and thus to reduce the Joule dissipation.

6.1.2 *The Influence on the Bubble Shape*

Let us first examine the bubble shape in the horizontal plane (Oz, Ox), as illustrated by three views from above together with color indications of the pressure variation (Fig. 6.5). Of course, without any MF the bubble section is circular (Fig. 6.5a). It then elongates in the Oz-direction, perpendicular to the horizontal MF (Fig. 6.5b), but this effect is reduced when the MF further increases. On the contrary, the bubble horizontal cross-section becomes again almost circular under a strong MF (Fig. 6.5c). The interface curvature providing such shape changes is the consequence of the pressure distribution outside the bubble, according to the relation

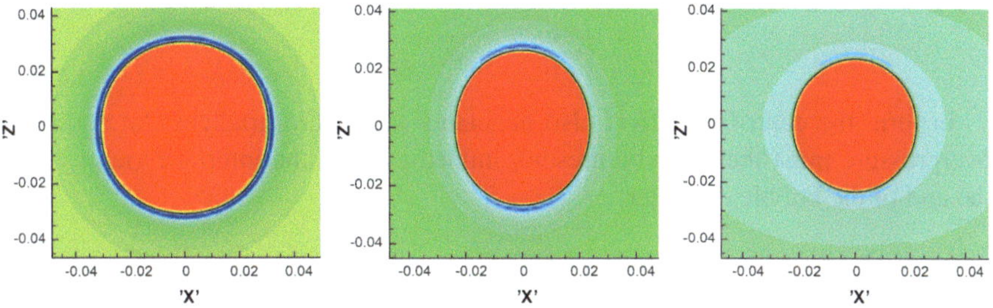

(a) $N = 0$ (b) $N = 3.6$ (c) $N = 24$

Fig. 6.5 View from above of the small bubble shape ($Eo = 2.2$) for different values of the interaction parameter, together with the illustration of the dimensionless pressure distribution. The magnetic field directs from left to right

$$p_{inside} - p_{outside} = \sigma\kappa + [\![\frac{B^2}{2\mu_m}]\!] \tag{6.1}$$

where μ_m is the termed magnetic permeability and $[\![\frac{B^2}{2\mu_m}]\!]$ represents the magnetic pressure jumps inside and outside of the bubble. Because in our problem both the magnetic field and the magnetic permeability are uniform, the magnetic pressure is the same inside and outside the bubble so that it does not play any role in the bubble shape. To interpret the observed surface changes, we must therefore understand the origin of the pressure variation around the bubble cross-section.

As seen in Sect. 6.1.1, the upward Lorentz force at both ends of the diameter in the Ox-direction is decreasing from its maximum value at the interface to zero over a typical distance a. It is balanced by an adverse vertical pressure gradient generating some overpressure at the top of the bubble, which can be estimated as: $p_1^* = j_z B a$ As a consequence of the Poisson equation to which the pressure obeys, the typical distance over which this overpressure decreases is larger than a. The equilibrium between these forces can only be global, the Lorentz force being predominant in the bubble vicinity, the pressure being predominant at distance a. This leads to a torque on a liquid metal domain surrounding the bubble in the vicinity of the Ox-direction. An eddy is therefore driven at each end of the Ox-diameter as shown in Fig. 6.4, without anything equivalent along the Oz-diameter. Neglecting viscosity, we easily get a first estimate of the linear velocity and angular velocity in this local vortex, given as

$$\begin{aligned} j_z B \cdot a &\approx \mu \frac{\omega a}{a^2} + \sigma_e u B \cdot Ba \\ &\approx \sigma_e \omega a \cdot B \cdot Ba \end{aligned} \tag{6.2}$$

where u is the linear velocity in the local vortex while ω is the angular vorticity, which is estimated as

$$\omega \approx \frac{j_z}{\sigma_e B a} \tag{6.3}$$

In turn, the centrifuge effect also generates a pressure gradient. An additional overpressure must therefore be present, immediately in the interface vicinity, near the end of the bubble diameter:

$$\begin{aligned} p_2^* &\approx \rho u^2 \\ &= \rho \omega^2 a^2 \\ &= \rho a^2 \frac{j_z^2}{\sigma_e^2 B^2 a^2} \\ &= \frac{\rho j_z^2}{\sigma_e^2 B^2} \end{aligned} \tag{6.4}$$

When comparing the two effects, we get:

$$\frac{p_2^*}{p_1^*} \approx \frac{\rho j_z}{\sigma_e^2 B^3 a} \tag{6.5}$$

It is then clear that, when the interaction parameter is moderate, the current density scaling as $j_z \approx \sigma_e B u$, this pressure ratio scales as N^{-1} and the centrifuge effect is predominant. The two vortices present on both sides of the bubble at the ends of the Ox-diameter, but not on the Oz-diameter (see Fig. 6.4), squeeze the bubble and force its elongation in the Oz-direction. Notice that this estimate cannot be valid in the limits $N \ll 1$, where j_z cancels and viscosity becomes significant, and $N \gg 1$, as discussed below. On the contrary, when $N \gg 1$ as already noticed, the current density must scale as $j_z \approx \rho u^2/Ba$ in order to maintain the equilibrium between pressure gradient and Lorentz force. As a consequence, the pressure ratio scales as N^{-2} and becomes smaller and smaller when N become very large. Therefore, the centrifuge effect disappears and the bubble recovers its circular shape.

This result may seem paradoxical and hard to reconcile with the well-established idea that a uniform MF tends to elongate in its direction any disturbance such as an eddy in turbulence (classic papers by Sommeria and Moreau [8], Davidson [2]). In turbulence, the source of this momentum diffusion in the MF direction (a reminiscence of Alfven waves when $Rm \ll 1$) is the fact that eddy currents loop with some part inside the eddy, where it brakes its vortical motion, and another part outside the eddy, where the current density has the opposite sign and, therefore, generates some vortical motion. This transfer of momentum and vorticity from the inside to the outside is not at all relevant in the case of a bubble: because the gas inside the bubble is insulating, the eddy currents loop totally outside and do not exist inside. As a consequence, there is no transfer of momentum from the inside of the disturbance (the bubble in our case, an eddy in turbulent flows) to the outside.

Now, consider the bubble shape in the vertical plane (Ox, Oy), illustrated by Fig. 6.6, together with the pressure distribution. These pictures show the shape of the

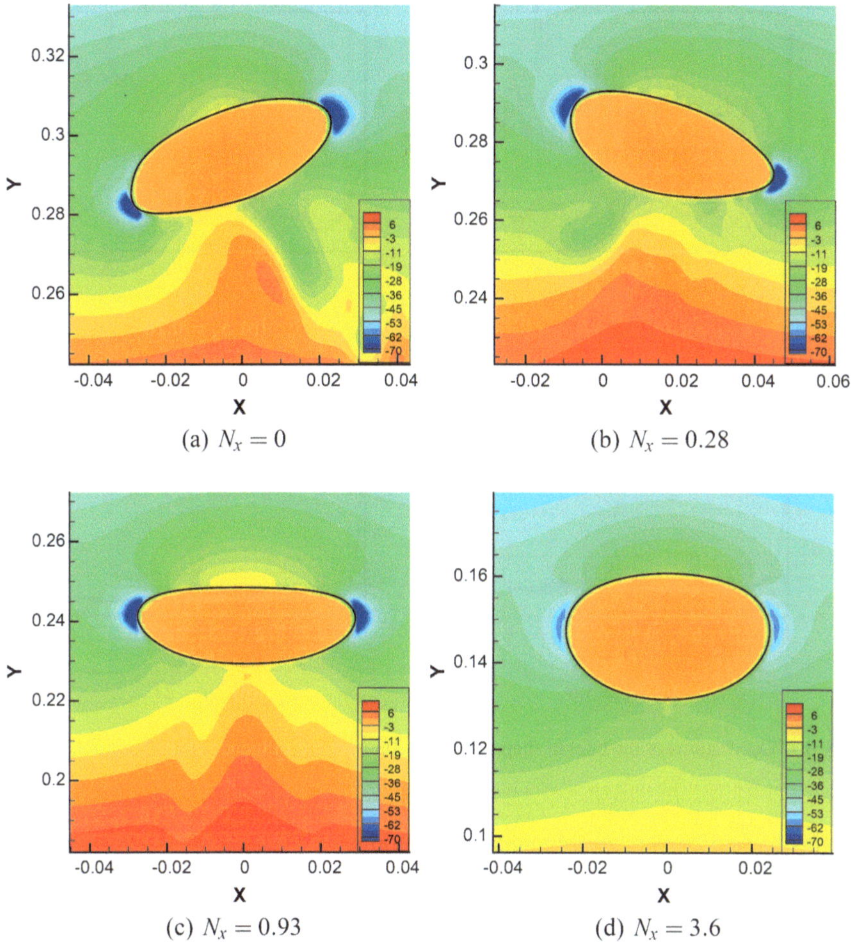

(a) $N_x = 0$ (b) $N_x = 0.28$

(c) $N_x = 0.93$ (d) $N_x = 3.6$

Fig. 6.6 Bubble shapes and pressure distributions in the vertical (Ox, Oy) plane, in the case of the small bubble $Eo = 2.2$, at dimensionless time $t^* = 15$, for different values of the horizontal MF. We can also observe that the wobbling motion of bubble is inhibited when $N_x > 1$. N_x refers to the horizontal magnetic field

small bubble ($Eo = 2.2$) at time $t^* = 15$. Clearly, when $N > 1$ the bubble motion appears more stable and the interface curvature at both ends of the Ox-diameter significantly smaller than when $N < 1$. Another issue of concern is the oscillation of the bubble shape during its rising, which can be also characterized by the axes d_1, d_2, d_3 which are illustrated in Chap. 5. Under the influence of the horizontal MF, the evolution of these axes is shown in Fig. 6.7 for different values of the interaction number. On Fig. 6.7a where $N = 0$, d_1 and d_2 elongate in the initial stage while d_3 is shortened, until a critical value of the aspect ratio $\chi = 3.6$, which is defined as the ratio of length between the longest and the shortest axes. Then the bubble shape

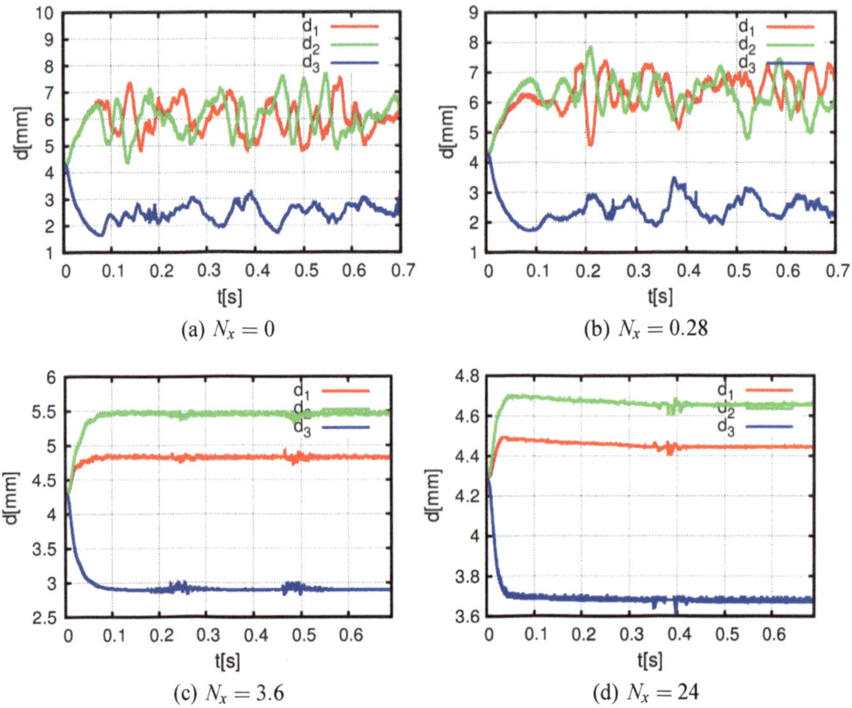

Fig. 6.7 Oscillations of the bubble shape ($Eo = 2.2$) under the influence of the horizontal MF

begins to oscillate, exhibiting small fluctuations. In case of $N = 0.28$, the shape oscillations seem not to be restrained by the horizontal MF and the value of the aspect ratio is almost as the same as without any MF. However, when the interaction number increases to $N = 3.6$, the shape oscillations disappear and the aspect ratio is also greatly reduced, as Fig. 6.7c shows. We should also notice that in the stable state, d_1 is shorter than d_2, in agreement with the anisotropic effect previously illustrated on Fig. 6.5c. In the last figure where $N = 24$, a smaller aspect ratio is observed and the difference between d_1 and d_2 is quite small. In summary, a strong MF results in small shape oscillations as well as a small aspect ratio, and this change in the interface shape is crucial to explain the suppression of the vortices [6, 7].

Finally, let us compare the influence of the horizontal MF (B_x) with that of the vertical MF (B_y), as shown on Fig. 6.8, still in the case of the small bubble ($Eo = 2.2$). The bubble flattening is larger in the presence of the horizontal MF than with the vertical MF. The overpressure on top and bottom of the bubble is also larger, leading to a smaller curvature of the interface near the top and bottom parts. This difference is a clear consequence of the direction of Lorentz force. When the magnetic field is horizontal, as shown in Fig. 6.2b, the Lorentz force acts downwards on the top surface of the bubble, pressing down and flattening the bubble. On the contrary, as shown in Chap. 5, under a vertical magnetic field, the Lorentz force acts nearly

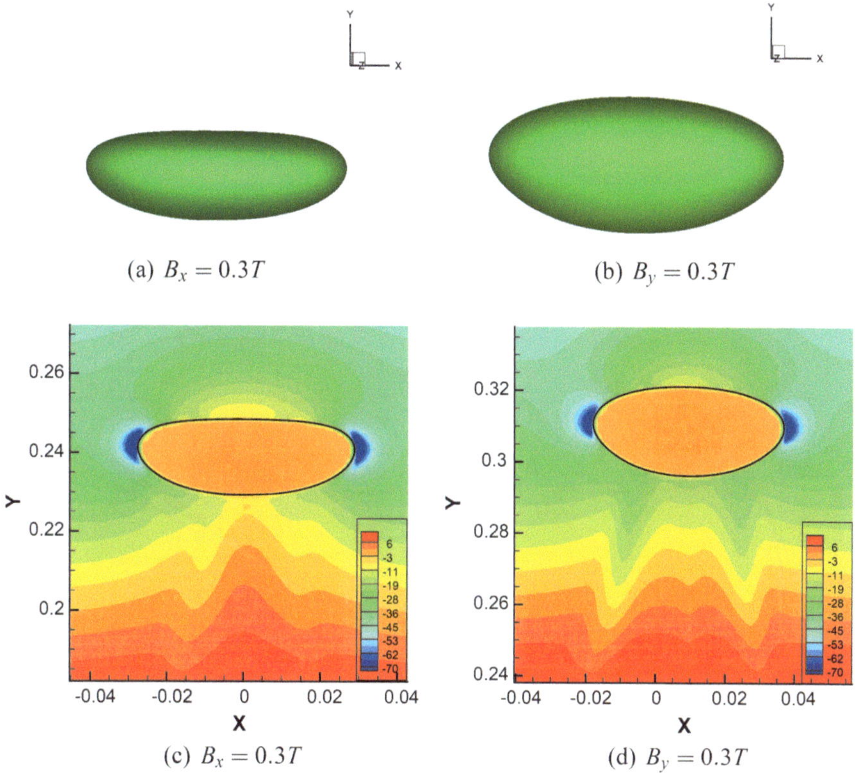

(a) $B_x = 0.3T$ (b) $B_y = 0.3T$

(c) $B_x = 0.3T$ (d) $B_y = 0.3T$

Fig. 6.8 Comparison between the influence of a horizontal (left hand side) and a vertical (right hand side) MF on the small bubble ($Eo = 2.2$). On the color scale, the pressure is given in dimensionless form

horizontally from around to the bubble center, so squeezing the bubble along the horizontal directions and making it more spherical.

6.1.3 The Influence on the Bubble Rising Velocity

Let us now investigate the influence of the horizontal MF on the bubble velocity. Figure 6.9 shows how the rising velocity and the rising height vary as time increases. Without any MF, for both bubbles, as soon as they are released, the rising velocity first increases almost without any fluctuations from its zero initial value to a maximum of the order of $U^* = 1.236$, obtained at time $t^* \simeq 5$, before oscillating and decreasing to the terminal value of $U^* = 1$. This result, which is expected with such a high Re number, is in agreement with the measurements by Shew et al. [7].

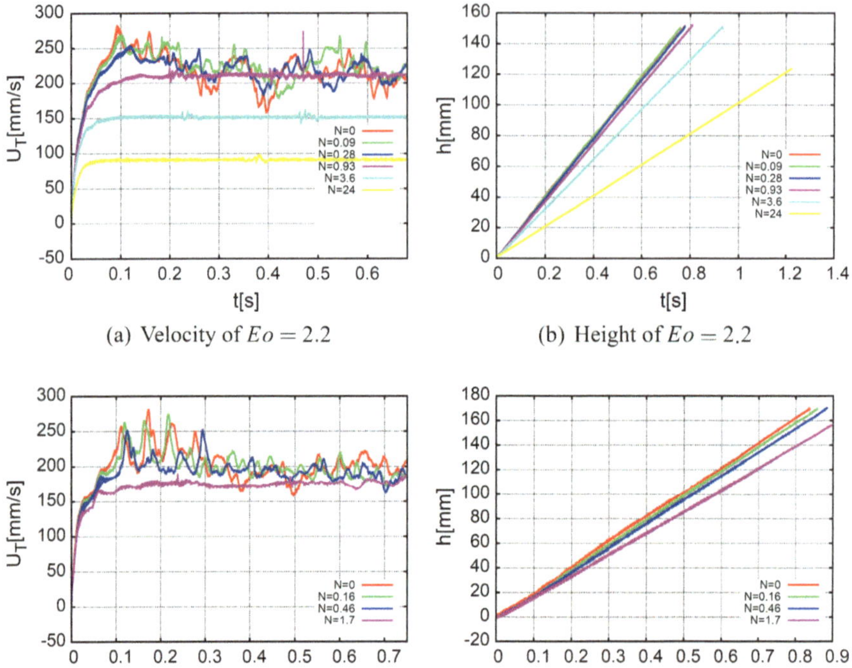

(a) Velocity of $Eo = 2.2$

(b) Height of $Eo = 2.2$

(c) Velocity of $Eo = 4.9$

(d) Height of $Eo = 4.9$

Fig. 6.9 Dimensionless rising velocity (left hand side: **a** and **c**) and bubble height (right hand side: **b** and **d**) versus time. The upper figures **a** and **b** correspond to the small bubble ($Eo = 2.2$), the lower figures **c** and **d** correspond to the large bubble ($Eo = 4.9$)

In the presence of a moderate horizontal MF ($0 < N < 0.28$), such that inertia is still significant, a similar behavior is observed for the small bubble ($Eo = 2.2$), with terminal velocity and height slightly above those obtained without any MF. When the MF is large enough ($N > 0.93$), on the contrary, the fluctuations are suppressed and the velocity increases monotonically with time until its terminal value, which is a decreasing function of the applied MF. For the large bubble ($Eo = 4.9$), with a still higher Re number, the fluctuations are more intense, and the mean bubble velocity and height monotonously decrease when the MF increases (Fig. 6.9c, d). These results are in fair agreement with the Mori et al. [5] measurements, but they disagree with the observations of Zhang et al. [10]. The reason of this disagreement is twofold: (i) in the Zhang experiments the MF is limited to 0.3 T, which is certainly a too small value to exhibit the asymptotic behavior under a high MF, (ii) their liquid melt (GaInSn) contains impurities which accumulate on the bubble interface and influence its behavior, which has been clarified in the previous chapter about the single bubble rising in vertical MF.

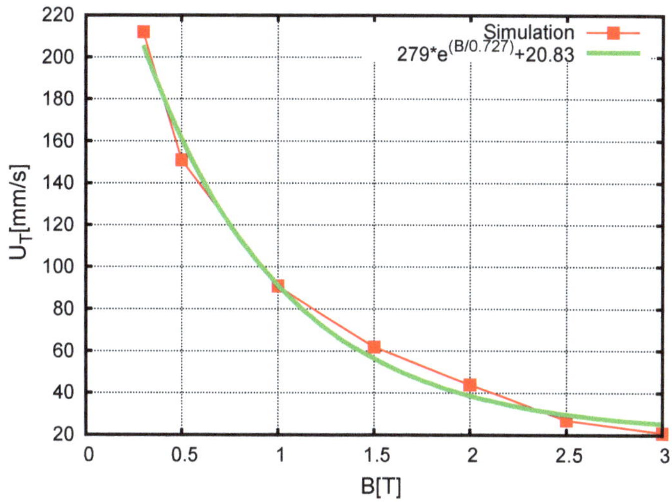

Fig. 6.10 Approximate exponential reduction of the bubble rising velocity when the intensity of the horizontal magnetic field increases, until a limit close to $U^* \approx 0.1$

To complement this analysis of the horizontal MF influence on the asymptotic rising velocity, we have extended our numerical simulation in the case of the small bubble ($Eo = 2.2$) to the MF value $B = 3\,\mathrm{T}$, corresponding to very high values of the interaction parameter (approximately 972). The results, presented on Fig. 6.10, fit rather well with an exponential decrease for the value above $U^* \approx 0.1$. They are in agreement with the measurements performed by Wang et al. [9]. Of course, they could not be found by Zhang et al. [10], whose experiments were limited to $B \leq 0.3\,\mathrm{T}$. The most challenging result may be this minimum for the rising velocity, which cannot be reduced below $U^* \approx 0.1$, whatever the strength of the horizontal magnetic field. When the interaction parameter gets very large, inertia becoming negligible, the Lorentz force must balance the pressure gradient, whose value has a necessary limit based on the hydrostatic pressure difference between the surfaces at the bubble top and bottom. Therefore, when B increases, the mean current density must decrease to maintain this force balance.

Again it is worth comparing these results on the rising velocity in the presence of a horizontal MF with those obtained in the presence of a vertical MF (Fig. 6.11). It should be noted that since both groups of simulations are conducted under the same magnetic intensity, but not the same interaction number, all figures are presented with B instead of N. This suggests the following remarks:

- Both bubbles have a velocity smaller with the horizontal MF than with the vertical one. This is clearly because the induced current and the associated braking mechanism are most efficient when u and B are generally perpendicular to each other.

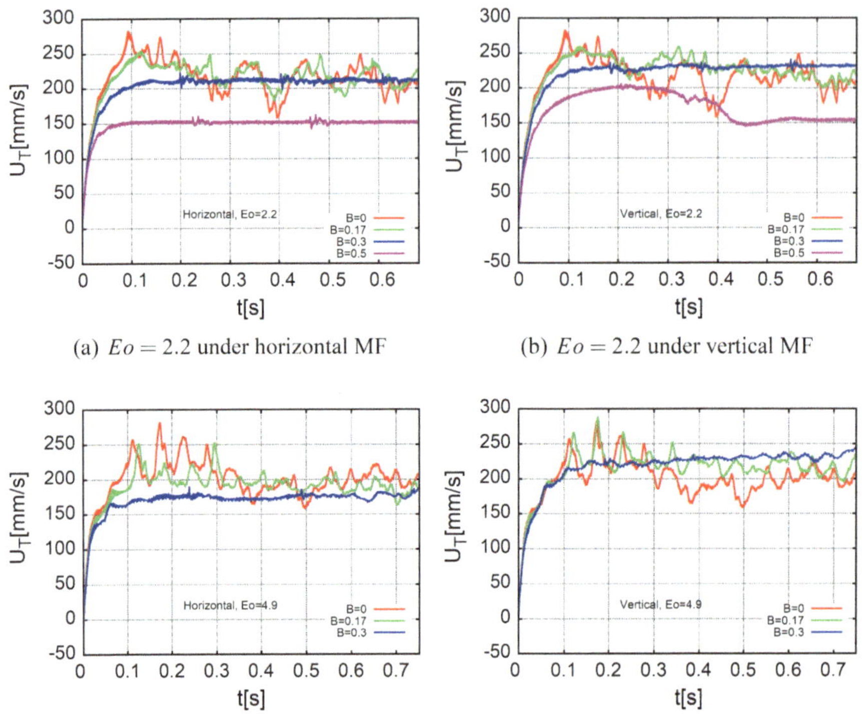

(a) $Eo = 2.2$ under horizontal MF

(b) $Eo = 2.2$ under vertical MF

(c) $Eo = 4.9$ under horizontal MF

(d) $Eo = 4.9$ under vertical MF

Fig. 6.11 Comparison between the influences of the horizontal MF (left hand side) and the vertical MF (right hand side) on the rising bubble velocity for a small bubble ($Eo = 2.2$, **a** and **b**) and for a large bubble ($Eo = 4.9$, **c** and **d**)

- The "second path instability" which is caused by the flow fluctuations around the bubble under vertical MF and has been discussed in Chap. 5 in details, is never present when the MF is horizontal. On the contrary, when B is large enough, the rising velocity reaches rapidly its asymptotic value and then remains constant.
- Both horizontal and vertical MF have a stabilizing influence on the velocity fluctuations, which are damped out when $B > 0.3$ T. This can be explained by the fact the three-dimensional fluctuations are systematically damped by the MF, whatever its direction.

This comparison between the influences of the horizontal and vertical MF directions is also illustrated on Fig. 6.12 in the case of the small bubble ($Eo = 2.2$) at time $t^* = 10$ with $B = 0.5$ T. It must be noticed that the wake behind the bubble is much narrower with the horizontal MF than with the vertical one, and that its intensity is also much smaller.

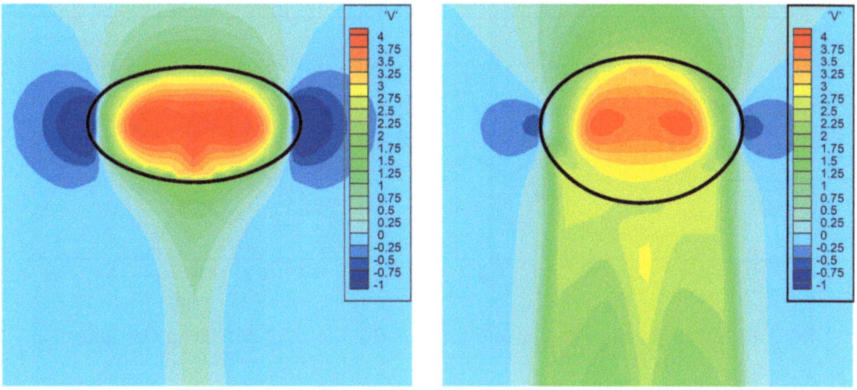

(a) Horizontal magnetic field (b) Vertical magnetic field

Fig. 6.12 Comparison of the influences of the horizontal MF and the vertical MF on the liquid velocity in the bubble wake for: $Eo = 2.2$, $B = 0.5\,\mathrm{T}$, $t^* = 10$. On the color scale, the velocity is given with dimensionless form

6.1.4 The Influence on the Bubble Trajectory

The fluctuating lateral drift of the bubble during its ascension through the liquid metal is shown on Fig. 6.13 both for the small bubble ($Eo = 2.2$, upper graphs (a) and (b)) and for the large one ($Eo = 4.9$, lower graphs (c) and (d)). The clearest effect is the total suppression of all horizontal oscillations of the bubble trajectory when $N > 0.93$ in the case of the small bubble, when $N > 1.7$ in the case of the large one. An analogous damping of these oscillations has been observed with a vertical MF, but several differences can be pointed out.

The "second path instability" does not exist when the MF is horizontal. Since we attribute this behavior to the flow instability in the liquid flow around the bubble, the question must be raised: is this turbulence suppressed by the horizontal MF? It has already been noticed (Sect. 6.1.1 and Fig. 6.2a) that the current density in the bubble vicinity concentrates near the diameter in the MF direction (Ox-direction), where its sign is such that the Lorentz force acts to push the liquid upwards and so to reduce its downward motion. This is consistent with the idea that the flow instability is restrained. Another argument leads to the same conclusion: since the bubble rising velocity is slower with the horizontal MF than with the vertical one, the Re number is also smaller and the liquid flow more stable.

Whereas the bubble rising is promoted by the vertical MF as shown in Chap. 5, it now appears that the horizontal MF makes the rising motion slower, even before the suppression of the path instabilities (see Fig. 6.9a, c). This can be attributed to the drag generated by the horizontal MF, whose influence is larger than the reduction of the path instabilities. In comparison, with a vertical MF making the bubble trajectory straighter, the rising velocity is larger, because of the reduction of the path instabilities.

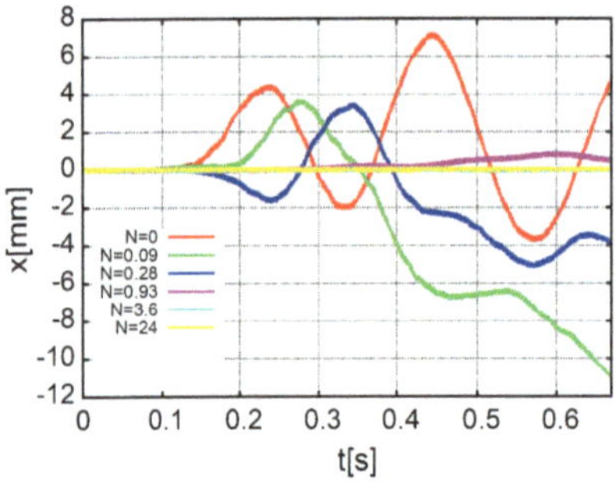

(a) $Eo = 2.2$ drift along $Ox-$ direction

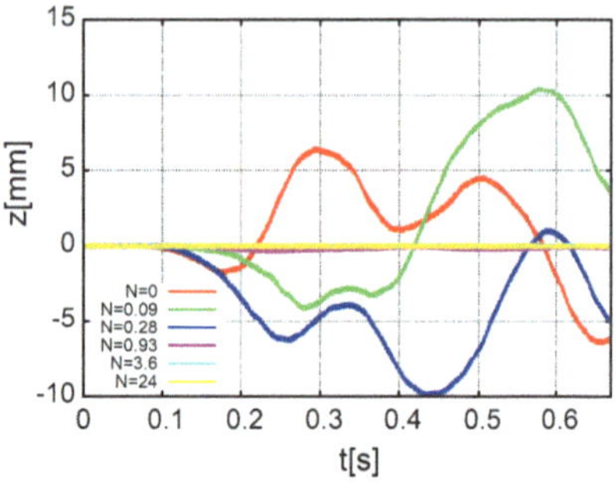

(b) $Eo = 2.2$ drift along $Oz-$ direction

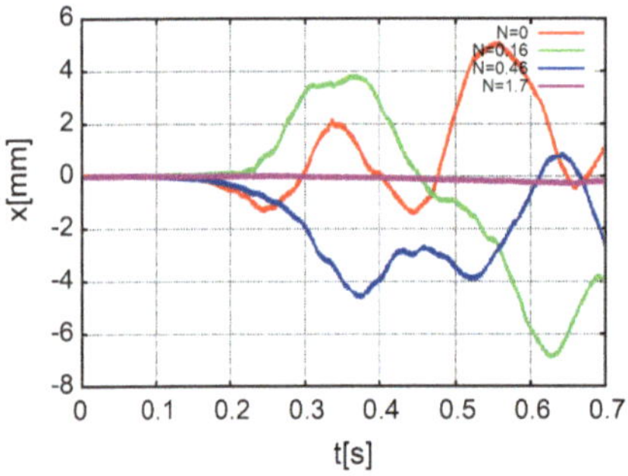

(c) $Eo = 4.9$ drift along $Ox-$ direction

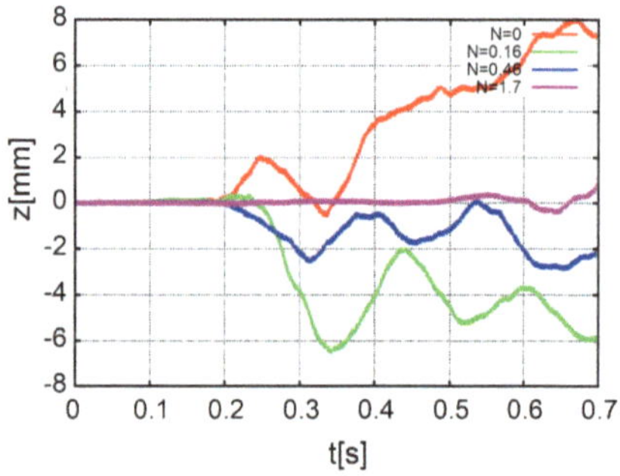

(d) $Eo = 4.9$ drift along $Oz-$ direction

Fig. 6.13 Typical horizontal deviations of the bubble trajectory versus time, both for the small bubble ($Eo = 2.2$, graphs **a** and **b**) and for the large one ($Eo = 4.9$, graphs **c** and **d**), on either the Ox-direction (graphs **a** and **c**) or the Oz-direction (graphs **b** and **d**)

6.1.5 The Influence on the Vortex Structures

Similarly, as widely known that the instability in the wake of bubble should be responsible for the path instability. Then, a particular attention must be paid to the evolutions of vortical structures under the influence of the horizontal MF.

The wake of the small bubble is found to have three typical organizations, which are presented on Fig. 6.14 at the same time $t^* = 15$, for different values of the horizontal MF, corresponding to $N = 0$, 0.28 and 0.93, respectively. In the figures, two values of ω_y are presented, with dark color denoting $\omega_y^* = -1$ and light color denoting $\omega_y^* = 1$. First, without any MF, the hairpin vortices, already observed in the experiments performed by Brücker [1] and numerically predicted by Gaudlitz and Adams [3], are present even with a larger Reynolds number. Second, these structures start to disappear when $N = 0.28$, and their twisting motion is destroyed. Finally, as soon as $N = 0.93$, in the quite stable wake, the vorticity becomes almost perfectly aligned in the vertical direction. This transformation of vortex structures are similar to that under vertical MF, however, when the lack of axial symmetry manifests

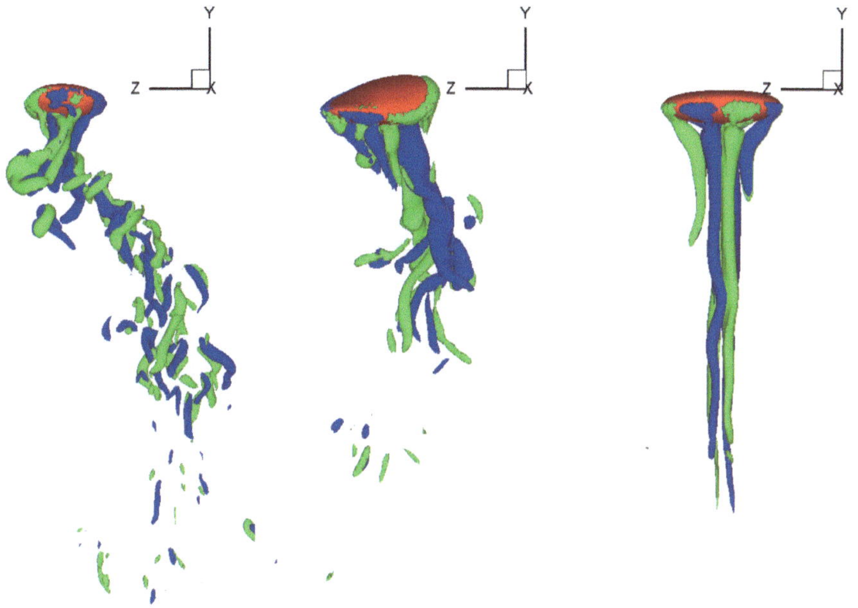

(a) $N = 0$ (b) $N = 0.28$ (c) $N = 0.93$

Fig. 6.14 Influence of the horizontal MF on the three-dimensional wake behind the small bubble ($Eo = 2.2$), for three values of the interaction parameter. The dark color corresponds to $\omega_y^* = -1$ while the light color corresponds to $\omega_y^* = 1$

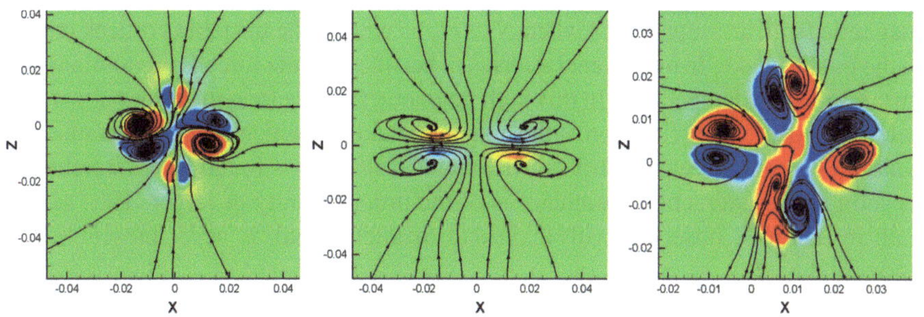

(a) $N_x = 0.93$ (b) $N_x = 3.6$ (c) $N_y = 0.872$

Fig. 6.15 Horizontal cross-sections of the small bubble ($Eo = 2.2$) wake, at $H^* = 0.93$ below the bubble bottom, under a horizontal MF (a and b for N_x) or a vertical MF (c for N_y). On the color scale, the magnitude of ω_y^* is given with dimensionless form. In all three cases the wake is stable

itself by the fact that the magnitude of the vorticity gets significantly smaller in the plane (Oy, Oz) perpendicular to the MF than in the (Ox, Oy) plane. Figure 6.15 illustrates this effect in horizontal $Ox - Oz$ cross-sections located $H^* = 0.93$ below the bubble bottom, where the color scale represents the magnitude of ω_y^*. It appears that this anisotropy, which is absent under a vertical MF $N_y = 0.872$ with eight almost periodic vortices (Fig. 6.15c), is already important under a horizontal MF when $N_x = 0.93$ and that it becomes still more pronounced when $N_x = 3.6$.

6.2 Summary

The influence of the horizontal MF on the motion of a single bubble rising in liquid GaInSn is investigated numerically in this chapter, showing a striking difference from that under the influence of a vertical MF.

A small magnetic field ($N < 1$) is sufficient to elongate the bubble shape in the direction perpendicular to the magnetic field. This elongation is due to the centrifuge overpressure generated by two vortices present on both sides of the bubble diameter in the magnetic field direction, which is squeezing the bubble. When $N \gg 1$, inertia and the centrifuge effect become negligible in comparison with the Lorentz force, essentially balanced by the pressure gradient, and the bubble recovers its circular form. This illustrates how the axial symmetry of the liquid metal flow around the bubble, previously noticed with a vertical magnetic field, is suppressed by the horizontal magnetic field. Other aspects of this anisotropy are exhibited, namely in the organization of the bubble wake.

By investigating the evolution of bubble shape under the influence of a horizontal magnetic field, smaller aspect ratio and shape oscillations are observed, which contribute to reduce the generation of vorticity in the bubble vicinity and correspondingly to make the bubble motion more stable.

The fluctuations of the bubble rising velocity, usually present when there is no magnetic field, are suppressed when $N \gg 1$, as well as the unstable vortices present in the bubble wake. In more general terms, when the Lorentz force is predominant, all instabilities are damped out. In this high magnetic field regime, the terminal velocity becomes a decreasing function of the magnetic field, for which an approximate exponential expression seems justified. The bubble trajectory is found to be straightened as soon as the interaction parameter is of the order of unity or larger. The "second path instability", observed with a vertical magnetic field, is never present with a horizontal field.

In the range of small or moderate magnetic fields, there is a fair agreement between our numerical results and the available experimental data by Zhang et al. [10], obtained with an Ar bubble rising through the liquid alloy GaInSn, in the presence of a magnetic field limited to 0.3 T. The remaining minor disagreements between our numerical results and this experiment seem to be due to the unavoidable presence of impurities in their melt, which act on the effective surface tension on the bubble interface. Our numerical results seem to be the only ones to provide the asymptotic bubble behavior when the interaction parameter gets very large.

References

1. Brücker C (1999) Structure and dynamics of the wake of bubbles and its relevance for bubble interaction. Phys Fluids 11(7):1781–1796
2. Davidson PA (1995) Magnetic damping of jets and vortices. J Fluid Mech 299:153–186
3. Gaudlitz D, Adams NA (2009) Numerical investigation of rising bubble wake and shape variations. Phys Fluids 21(12):122102
4. Miao X, Lucas D, Ren Z et al (2013) Numerical modeling of bubble-driven liquid metal flows with external static magnetic field. Int J Multiph Flow 48:32–45
5. Mori Y, Hijikata K, Kuriyama I (1977) Experimental study of bubble motion in mercury with and without a magnetic field. J Heat Transf 99(3):404–410
6. Mougin G, Magnaudet J (2001) Path instability of a rising bubble. Phys Rev Lett 88(1):014502
7. Shew WL, Pinton JF (2006) Dynamical model of bubble path instability. Phys Rev Lett 97(14):144508
8. Sommeria JL, Moreau R (1982) Why, how, and when, MHD turbulence becomes two-dimensional. J Fluid Mech 118:507–518
9. Wang ZH, Wang SD, Meng X et al (2017) UDV measurements of single bubble rising in a liquid metal Galinstan with a transverse magnetic field. Int J Multiph Flow 94:201–208
10. Zhang C, Eckert S, Gerbeth G (2005) Experimental study of single bubble motion in a liquid metal column exposed to a DC magnetic field. Int J Multiph Flow 31(7):824–842
11. Zhang C, Eckert S, Gerbeth G (2007) The flow structure of a bubble-driven liquid-metal jet in a horizontal magnetic field. J Fluid Mech 575:57–82

Chapter 7
The Path Transition from Zigzag to Spiral During the Single Bubble Motion

In the previous two chapters (Chaps. 5 and 6), we confirm that the bubble motion is greatly influenced by imposing an external MF, which modifies the bubble shapes, the vortex structures, and also the rising paths. Meanwhile, the relations between the wake evolution and the path transition are more clear by investigating the transition process under the influence of the MF. However, the physical mechanics causing the path instability are still not so clear. In this chapter, we try to study a particular case of the vortex evolutions when the rising path of the single bubble transits from zigzag to spiral.

7.1 Introduction

As introduced in Sect. 5.1 of Chap. 5, different vortex structures are observed behind the single bubble when it travels in a zigzag or spiral path, respectively in the experimental or numerical studies. Typical experimental results are presented by Lunde and Perskin [6], Brücker [2], Ellingsen and Risso [5], and Vries et al. [4], while corresponding numerical results are also shown by Mougin and Magnaudet [9], Shew and Pinton [10], Cano-Lozano et al. [3] and Tripathi et al. [12]. All the results reveal that the double-threaded vortices are shed periodically in the zigzag stage while the vortex structures are more stable in the spiral stage, indicating that the pattern of the vortex structures are closely related with the rising path of the bubble.

Nevertheless, although the relation between the wake instability and the path instability is already recognized, it is still not clear why the bubble is sometimes observed to transit from zigzag to a spiral. Furthermore, although it is believed that the accumulation of vorticities on the bubble surface will lead to the vortex shedding, however, there is still no quantitative evidence about this. In the present study, we will answer the above questions by presenting the evolutions of the vortex structures during the zigzag and spiral motions, respectively. According to the experimental

J. Zhang, *The Developments and the Applications of the Numerical Algorithms in Simulating the Incompressible Magnetohydrodynamics with Complex Boundaries and Free Surfaces*, Springer Theses, https://doi.org/10.1007/978-981-10-6340-4_7

results by Zenit and Magnaudet [14], the path instability will even appear within low *Re* flows as long as the aspect ratio of the bubble exceeds a critical value. In the low *Re* regime, the wake structures are thought to be more elemental and regular than those under higher *Re* flows, where the vortex structures would be much more complicated. In addition, high *Re* flows contain more flow instabilities that it is difficult to identify the physical mechanism for the path transition. It is also pointed out by Cano Lozano et al. [3] that to simulate the high *Re* flows, capturing the boundary layer requires much thinner grids close to the bubble, and it is really time consuming or otherwise the results are not reliable. As a consequence, the path instability, together with the wake instability, will be studied within lower *Re* flows in the present chapter. The mechanics could be also used to understand the path transition from zigzag to spiral in high *Re* flows.

In particular, as presented in Chaps. 5 and 6 where the external MFs are imposed, the double-threaded vortex structures are influenced markedly, therefore, it can be used to further identify the relations between the unstable path patterns and the vortex structures. After understanding this, we will further validate that it is rather the vortex structures, especially the relative strength between the double vortex threads, that determine different path patterns of the rising bubble.

In this chapter, if we substitute the characteristic length of $L_0 = R$ and the characteristic velocity of $u_0 = \sqrt{gR}$ into Eq. 2.14, the Navier-Stokes equations take the variant form of

$$\frac{\partial \boldsymbol{u}^*}{\partial t} + \boldsymbol{u}^* \cdot \nabla \boldsymbol{u}^* = -\nabla p^* + \frac{1}{Ga}(\mu^*(\nabla \boldsymbol{u}^* + \nabla \boldsymbol{u}^{*T}) + \frac{1}{Bo}\kappa^* \delta_s^* \boldsymbol{n} + (\rho^* - 1)\boldsymbol{g}^* \tag{7.1}$$

Nevertheless, different from Eq. 2.14 which describes the single-phase flows, some new characteristic variables are used in the present equation, those are

$$\rho^* = \frac{\rho}{\rho_l},\ \mu^* = \frac{\mu}{\mu_l},\ \boldsymbol{g}^* = \frac{\boldsymbol{g}}{\|\boldsymbol{g}\|}$$

Therefore, the dynamic behaviors of the rising bubble are actually depended on four dimensionless parameters, which are ρ^*, μ^* and the other two are defined as:

$$Ga = \frac{\rho_l \sqrt{gR}R}{\mu_l},\ Bo = \frac{\rho_{lgR}^2}{\sigma}$$

It is obvious that the Galilei number (Ga) and the Bond number (Bo) are the variant forms of the Reynolds number (Re) and Weber number (We) respectively by replacing $\sqrt{gR}$ with u_0.

In this chapter, the density ratio and viscosity ratio between the gas phase and the liquid phase, namely ρ^* and μ^*, are fixed at 10^{-3} and 10^{-2} respectively. Therefore, the bubble motions are entirely characterized by Ga and Bo according to Eq. 7.1. When magnetic field is imposed, similarly, the dimensionless interaction number takes the form of $N = \frac{\sigma_e B^2 R}{\rho\sqrt{gR}}$ by using the characteristic velocity.

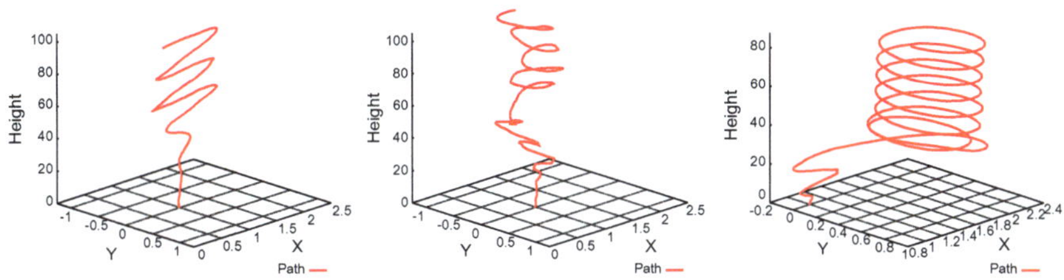

Fig. 7.1 Different paths of the single bubble motion ($Ga = 50$, $\rho^* = 1/1000$, $\mu^* = 1/100$). From left to right: $Bo = 1$ for zigzag, $Bo = 2.0$ for zigzag-spiral transition and $Bo = 3.0$ for spiral

7.2 The Zigzag and Spiral Motions of the Rising Bubble

Different flow conditions are investigated with a limited regime of Ga and Bo, which are varied between $35.36 \sim 70.72$ and $0.7 \sim 3$, respectively. However, in order to make this chapter more compact, we will focus on the flow condition of $Ga = 50$ while Bo is varied between 1.0, 2.0 and 3.0 in the following study, because we find that they come to the similar conclusions when the bubble travels within a zigzag or spiral path. Under $Ga = 50$, the terminal Re approaches 150, and the rising trajectories of the bubble centroid, with a three-dimensional view, are given in Fig. 7.1. As shown in the figures, different rising paths are observed as Bo increases, from an absolutely zigzag path ($Bo = 1.0$) to a zigzag/spiral transition path ($Bo = 2.0$) and finally being a perfect spiral path ($Bo = 3.0$).

It should be noted that our results are fully in accordance with the numerical results reported by Cano-Lozano et al. [3], who draw a phase diagram summarising different path patterns observed in their simulations, as presented in Fig. 22 of that paper. According to that diagram, the bubble motions in the present study (the parameters become $\{Ga, Bo\} = \{141.42, 4\}, \{141.42, 8\}, \{141.42, 12\}$ if $D = 2R$ is used as characteristic length) should fall into the planar zigzag, flattened spiral and spiral regions respectively.

7.2.1 *Zigzag Motion*

As shown in Fig. 7.1a, the bubble of $Bo = 1$ will perform a perfect zigzag motion, however, the projection of the path is not aligned either with x-axis or y-axis. In order to make things clearer, a coordinate transformation is applied on the original system to make the new x'-axis along with the zigzag direction. After that, the rising velocity, the transverse position and the transverse velocity of the bubble centroid are drawn in Fig. 7.2 based on the transformed coordinate. From Fig. 7.2b, c, it is clear that the bubble travels in a zigzag path.

As shown in Fig. 7.2a, after an acceleration process until $t^* = 10$, the rising velocity of the bubble will fluctuate around $V^* = 1.6$. Such velocity fluctuation

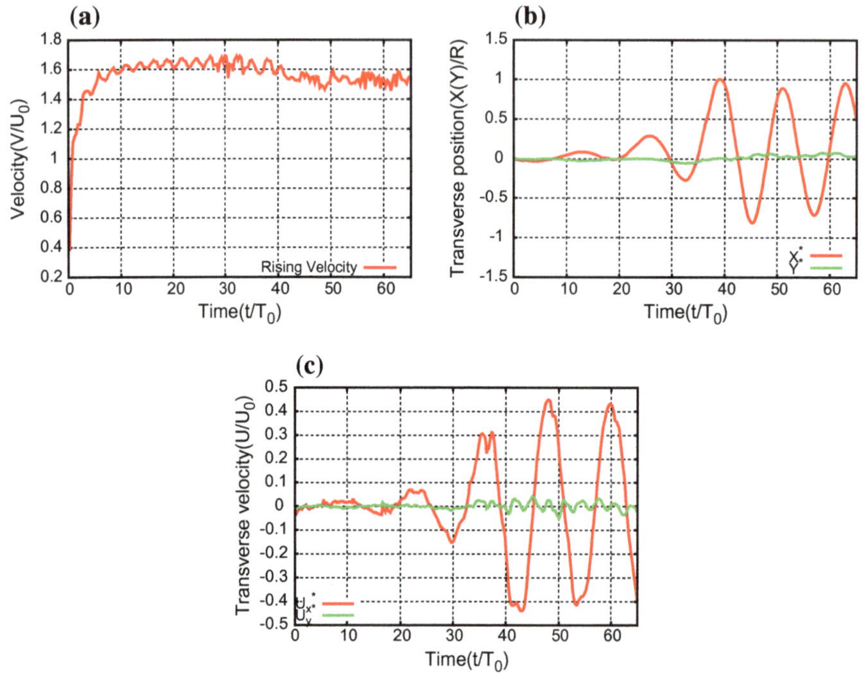

Fig. 7.2 **a** The rising velocity of the zigzag bubble under $Ga = 50$ and $Bo = 1$. **b** The transverse position of bubble centroid after coordinate transformation. **c** The transverse velocity of bubble centroid after coordinate transformation. It is observed that the rising velocity suddenly drops at $t^* \approx 39$, while perfect zigzag motion is observed after $t^* > 30$

is probably related to the shape oscillations according to the study by Lunde and Peskin [7] and a detailed analysis will be given in the next paragraph. Following the fluctuation, an abrupt dropping of rising velocity is observed at $t^* \approx 40$ when the bubble reaches its furthermost position at that moment, as shown in Fig. 7.2b. To find the physical mechanism inducing such type of rising instability, the vortex structures are shown in Fig. 7.3, where the streamwise vorticity of ω_z^* is present at $t^* = 39$ and $t^* = 40$ respectively, with dark color indicating $\omega_z^* = -0.5$ and light color indicating $\omega_z^* = 0.5$. In Fig. 7.3a, it is observed that vortex pair *I* are still attached to the bottom of the bubble, however in Fig. 7.3b, the vortex pair *I* are shed from the bubble. Therefore, the sudden drop of rising velocity should be attributed to the shedding of the vortex pairs, which will generate a drag force on the bubble motion.

Concerning the dynamics of the vortex behaviors during the zigzag motion, a detailed analysis will be discussed that four typical time instants are selected in a complete zigzag cycle, which is shown in Fig. 7.2b, respectively with $t^* = 39$, $t^* = 42$, $t^* = 45$ and $t^* = 48$ when the bubble passes the averaged positions and the furthermost positions. The positions of the bubble, as well as the structures of ω_z^* at the four typical time instants, are present in Fig. 7.4, where dark color corresponds

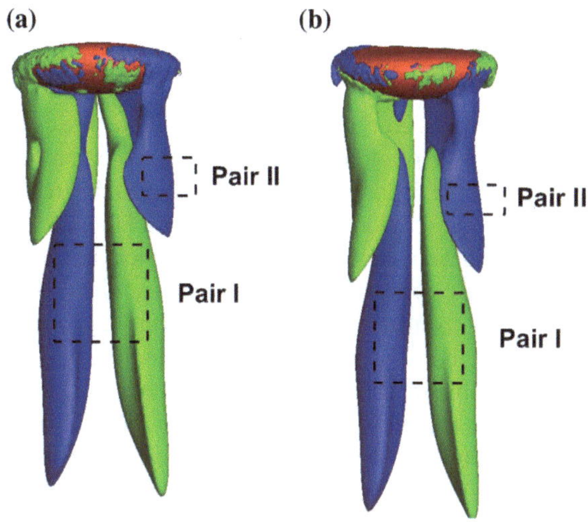

Fig. 7.3 The vortex structure of ω_z^* behind the bubble respectively at **a** $t^* = 39$ and **b** $t^* = 40$. It is observed that vortex pair *I* still attached at the bottom of the bubble at $t^* = 39$ while they are already shed at $t^* = 40$. Correspondingly, the rising velocity of the bubble, as shown in Fig. 7.2a, is observed to drop suddenly

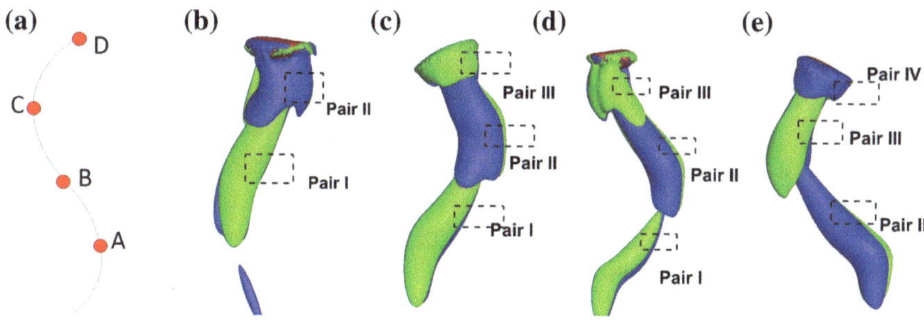

Fig. 7.4 The streamwise vortex structures behind the bubbles at different times in a zigzag period. **a** the typical positions at different times, **b** $t^* = 39$, **c** $t^* = 42$, **d** $t^* = 45$ and e $t^* = 48$. The light color corresponds to $\omega_z^* = 0.5$ while the dark color corresponds to $\omega_z^* = -0.5$. It is observed that the double threaded wake attached at the bubble rear and their signs are changed twice in a zigzag period

to $\omega_z^* = -0.5$ and light color corresponds to $\omega_z^* = 0.5$ respectively. In the figures, the double threaded wakes (one thread blocks the view of the other in the figures) are observed to attach at the bubble rear during the zigzag motion, and this pattern of vortex structure is also reported by various numerical and experimental studies, as shown by Mougin [9] and Veris [4]. Another similarity between our numerical solutions and their results lies in the observation that the vortex filaments will change its sign twice during a zigzag cycle, as shown by comparing Fig. 7.4b with d.

When the bubble passes the furthermost position at $t^* = 39$, vortex *Pair I* just complete the shedding process while *Pair II* are shedding at the right side of the bubble. Meanwhile, new vortex pairs, which signs are opposite to *Pair II* are accumulated on the bubble surface. At $t^* = 42$ when the bubble passes the average position, *Pair II* have developed to a longer thread and the accumulated new vortices, *Pair III*, begin to shed. At the following stage, when the bubble reaches the opposite furthermost position at $t^* = 45$, vortex *Pair III* are shedding at the left side of the bubble. This process is the same as that in Fig. 7.4b. When the bubble returns to the average position again at $t^* = 48$, *Pair III* begin to detach from the bubble, and the new vortex *Pair IV* begin to shed from the bubble surface again.

Therefore, the dynamic behaviors of the wakes behind the bubble can be summarized as: (a) when the bubble reaches the furthermost positions, the shedding vortex pairs (*Pair II* in Fig. 7.4b and *Pair III* in Fig. 7.4d) at the bubble rear will produce a transverse vortical force to drive the bubble taking reverse motion. To balance this force, the bubble will rotate counterclockwise (resp. clockwise) to generate a transverse component of the buoyancy force to partly compensate the vortical force. During this period, new vorticities begin to gather at the lateral surface of the bubble. (b) when the bubble passes the average position, the amount of the accumulated vorticities on the bubble surface (*Pair III* in Fig. 7.4c and *Pair IV* in Fig. 7.4e) reaches the critical value and a new round of vortex shedding process starts. However, the new vortex pairs, with opposite signs to the former pairs, will result in an opposite transverse force to decelerate the bubble while the drift angle is also reduced.

The mechanism to generate such a vortical force by the vortex pairs is shown in Fig. 7.5a, which presents the top view of the counter-rotating vortices of Fig. 7.4b. In the figure, the black solid line is the bubble interface while the dotted line is the double wake threads attached to the bubble bottom. According to the theory of vortex pairs [1], when the vortex strengths of the two vortex filaments are equal, a horizontal velocity directing rightward will be induced, and under the conservation law of momentum, the bubble will be pushed leftwards. This theoretical deduction is consistent with the motion direction of the numerical result. When the bubble passes the average position at $t^* = 42$, the newborn vortex *Pair III* of opposite signs will induce an opposite velocity, which acts to decelerate the bubble motion. Nevertheless, it should be noted that to induce such a straight-line velocity without angular velocity, the vortex strength between the counter-rotating vortices should be equal. To quantitatively prove this, a cross plane of $4R$ downstream the bubble bottom is cut out, then the vortex isosurfaces on that plane are shown in Fig. 7.5b while the red color denotes $\omega_z^* = 0.5$ and the blue color is $\omega_z^* = -0.5$. It is observed that the double vortex threads are well matched in strength. Moreover, through computation, the time history of the arithmetic sum of streamwise vorticities, being $\tau_{plane} = \int \omega_z dS$ (much different from $\| \omega_z \|$) on that plane is also shown in Fig. 7.5c, in which the value of zero indicates that within a zigzag period, the positive and the negative vortices are symmetric.

In order to demonstrate that the zigzag motion is induced by the periodic shedding of vortex pairs, the norm of the vorticities accumulated on bubble surfaces, namely $\| \tau \| = \int_S \| \omega \| dS$, are plotted in Fig. 7.6 versus time. Fig. 7.6a is the amount of

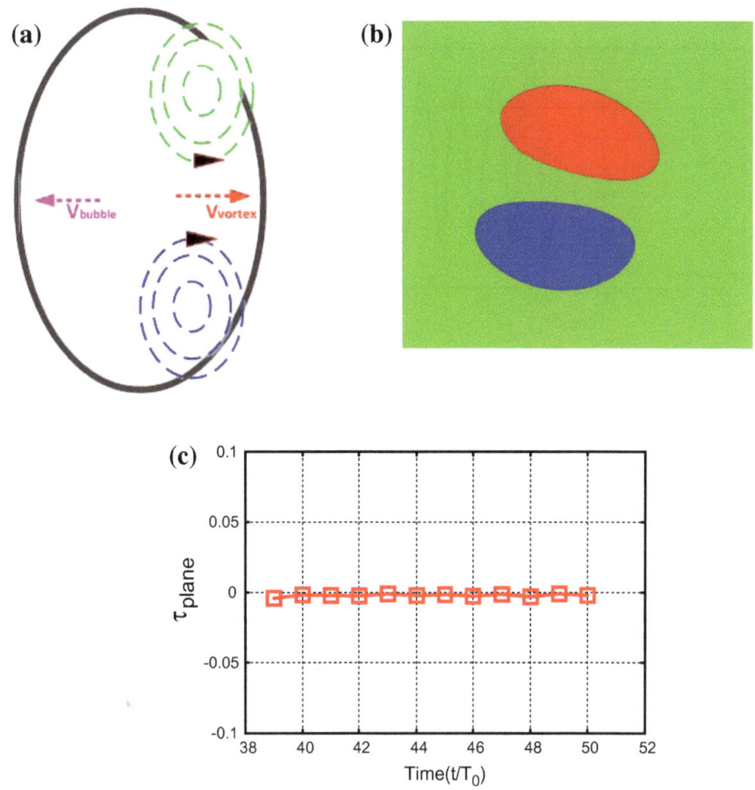

Fig. 7.5 The information of the vortex pair in the zigzag motion **a** the sketch of the vortex pairs. When the two threads are equal in strength, a horizontal velocity directing rightward is induced, under the effect of reacting force, the bubble will be pushed leftwards. **b** the contour of the vortex pairs at the plane of $4R$ downstream the bubble bottom, with red color denoting $\omega_z^* = 0.5$ and blue color denoting $\omega_z^* = -0.5$. It is observed that they are well matched in strength. **c** The time series of the arithmetic sum of the streamwise vorticities intergraded on the cut plane as $\tau_{plane} = \int \omega_z dS$, proving that the counter-rotating vortices are perfectly symmetric

the streamwise component $\| \omega_z^* \|$ while Fig. 7.6b is the total vorticities $\| \omega^* \|$. From Fig. 7.6a, periodic oscillations are observed during the zigzag motion, and there is a strong interaction between the oscillatory path and the accumulation of $\| \omega_z^* \|$ at the bubble surface. Whenever the bubble passes the furthermost positions, such that at $t^* = 39$, $t^* = 45$, $t^* = 51$ and $t^* = 57$, the accumulated $\| \omega_z^* \|$ on bubble surface decreases to the minimize value, indicating that the vortex pairs are completely shed from the bubble surface. This is in accordance with what we present in the above paragraph, as shown in Fig. 7.4b, d. By contrary, once the bubble passes the average position, such that at $t^* = 42$, $t^* = 48$, $t^* = 54$ and $t^* = 60$, the amount of $\| \omega_z^* \|$ on bubble surface reaches its maximum value, indicating that the accumulation process of vorticities is completed and the shedding process is going to start, this is also

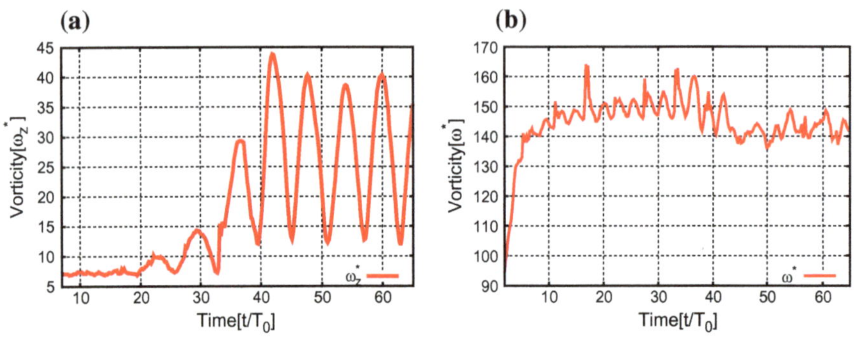

Fig. 7.6 Time histories of the norm of the vorticities accumulated on the bubble surface **a** the streamwise component $\| \omega_z^* \|$. **b** the total vorticity $\| \omega^* \|$. It is observed that the oscillation frequency of the amounts of $\| \omega_z^* \|$ is twice that of the zigzag path, indicating that the shedding process of ω_z^* induces the oscillation of zigzag path. However, there is not direct relationship between the amounts of $\| \omega^* \|$ and the zigzag path

consistent with that shown in Fig. 7.4c, e. However, in Fig. 7.6b, where the time history of $\| \omega^* \|$ is present, the variation trend is very different, and it seems that the zigzag motion has little relationship with that of $\| \omega^* \|$ accumulated on bubble surface. In fact, the variation of $\| \omega^* \|$ seems to be in accordance with that of the rising velocity, as shown in Fig. 7.2a. Therefore, it is actually $\| \omega_z^* \|$ accumulated on the bubble surface to trigger the zigzag motion. When the amount of $\| \omega_z^* \|$ accumulated on the bubble surface exceeds a critical value, a zigzag motion will be induced by the shedding of the vorticities.

Magnaudet and Mougin [8] also give the similar interpretation that they found if the maximum ω on the bubble surface reached the critical value ω_{cr}, the wake instability will take place. They also point out that this mechanism is independent of the boundary condition at the interface, irrespective of the solid sphere or the rising bubble. As a consequence, the time histories of the maximum value of $\| \omega_z^* \|$ and $\| \omega^* \|$ on the bubble surface are present in Fig. 7.7. From the figure, similar variation trends are observed compared to that of the total amounts of vorticties accumulated on the bubble surface, which also indicates that the zigzag path instability is more likely to be depended on ω_z^* instead of ω^*.

Another important characteristic of the rising bubble is the shape oscillations during the unstable motion. As found by Lunde and Peskin [7] and Veldhuis et al. [13], there is a strong interaction between the shape oscillations of mode (2,0) and the bubble motion, therefore we are inspired to check whether the fluctuation of the rising velocity in Fig. 7.2a is related to the shape oscillations of the bubble.

The evolution of the aspect ratio χ, which is defined as the ratio between the major diameters and the minor diameter of the bubble, is used to study the shape effect. The sketch of the bubble shape is shown in Fig. 7.8a that χ is defined as $\chi_1 = d_1/d_3$ and

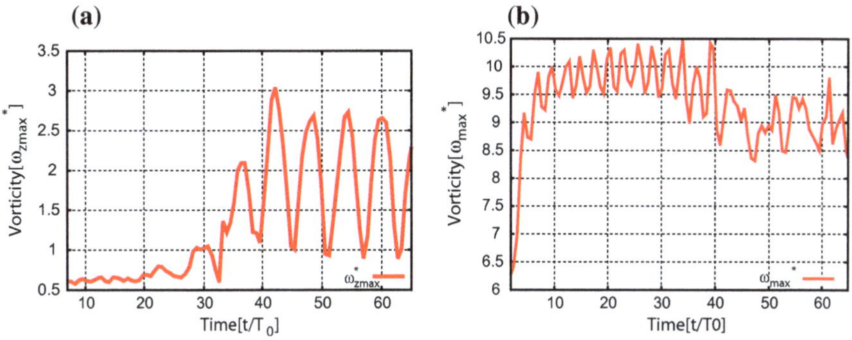

Fig. 7.7 Time histories of the maximum value of the vorticties accumulated on the bubble surface **a** the streamwise component $\| \omega_z^* \|$. **b** the total vorticity vector $\| \omega^* \|$. The similar variation trend is observed compared with that in Fig. 7.6

$\chi_2 = d_2/d_3$. Lunde and Perkins [7] suggested that bubbles under symmetric shape oscillations undergo a dominant mode, which can appear in two different forms. The (2,0) mode represents the bubble shape oscillates between the oblate and prolate, based on the evolution process of $\sqrt{\chi_1 \chi_2}$. For the other (2,2) mode, the bubble rotates about its minor axis and this mode can be identified with the evolution of χ_1/χ_2. In Fig. 7.8b, the criterion for both shape oscillation modes are present, it is observed that before the zigzag motion taking place ($t^* < 30$), the (2,0) mode shows a high frequency of $f^* = 0.36$ while no noticeable oscillation is observed in (2,2) mode. However, after the zigzag motion happens ($t^* > 30$), both the (2,0) mode and the (2,2) mode follow the shedding frequency of the streamwise voticities, by comparing with Fig. 7.6a. This result is quite consistent with the experimental observation by Brücker [2], who found that the oscillation period of the bubble breadth was just half of the zigzag period. This periodic oscillation of the bubble shape is caused by the uneven pressure distribution in the equatorial plane of the bubble when it arrives at the inversion points.

Besides, Lunde and Peskin [7] think the oscillations of the bubble shape reflect the fluctuations in the added mass coefficient, which leads to the fluctuated motion of the bubble. This is illustrated in Fig. 7.8c by showing the acceleration of the bubble centroid versus time. It should be noted that the time evolution of the bubble velocity is computed as $U(t) = \sqrt{u(t)^2 + v(t)^2 + w(t)^2}$, and then the acceleration of the bubble $A(t) = dU(t)/dt$ can be determined. As shown in Fig. 7.8c, the oscillations of the bubble motion exhibits a frequency of $f^* = 0.36$ and $f^* = 0.40$ respectively before and after the zigzag motion happens. The results indicate that there is a strong direct interaction between shape oscillation of mode (2,0) and the motion of the bubble.

Moreover, they also give a simple mode of shape oscillations [7] that the frequency of the mode(2,0) can be obtained by

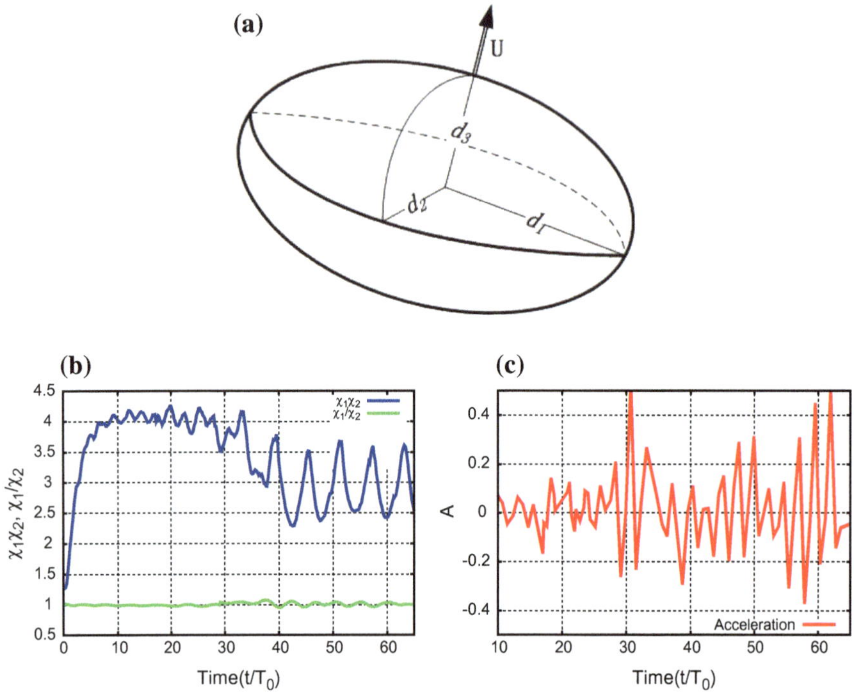

Fig. 7.8 **a** Sketch of the bubble shape, while aspect ratio is defined as $\chi_1 = d_1/d_3$ and $\chi_2 = d_2/d_3$. **b** The time series of the shape oscillation: (2,0) mode criterion $\sqrt{\chi_1\chi_2}$ and (2,2) mode criterion χ_1/χ_2. **c** The time series of the acceleration of the bubble

$$f_{2,0} = \frac{1}{2\pi}\sqrt{\frac{16\sqrt{2}\chi^2\sigma}{\rho_1(\chi^2+1)^{3/2}R^3}} \tag{7.2}$$

A theoretical frequency of $f^*_{theo} = 0.362$ is calculated by substituting the parameters into Eq. 7.2, and the numerical result of $f^*_{num} = 0.36$, as shown above, is in good agreement with this theoretical value.

7.2.2 Path Transition from Zigzag to Spiral

In Fig. 7.1b, the rising path of the bubble is observed to transit from zigzag to spiral in case of $Bo = 2.0$. This is more obvious as shown in Fig. 7.9b, where the position of the bubble centroid along x-direction and y-direction are present respectively. From the figure, it is observed that when $t^* < 40$, the bubble travels within a zigzag path, which turns to be a spiral one when $t^* > 40$. The time series of the rising velocity and the transverse velocities are also plotted in Fig. 7.9a, c, respectively. It is observed

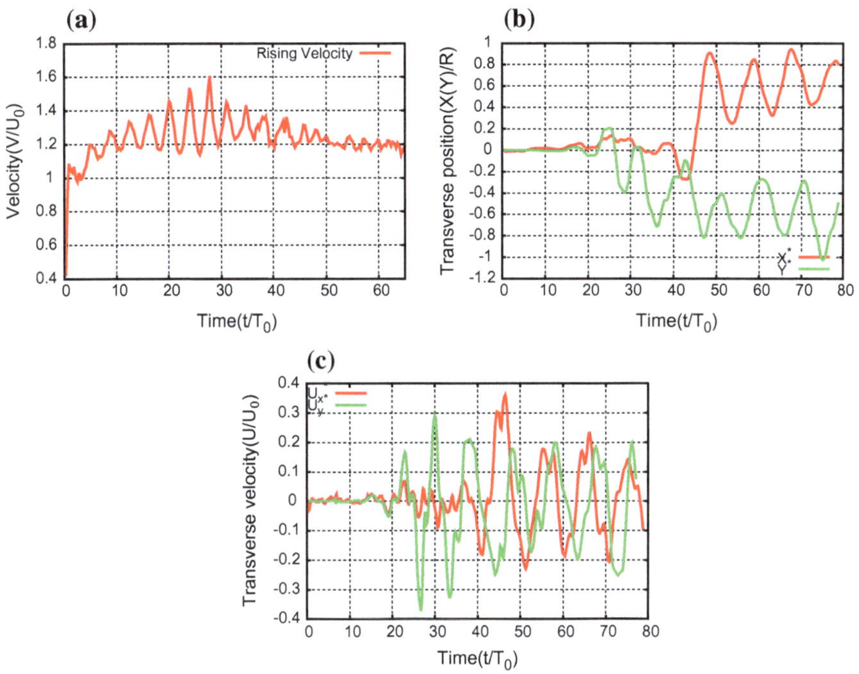

Fig. 7.9 **a** The rising velocity of the zigzag/spiral bubble under $Ga = 50$ and $Bo = 2.0$. **b** The time evolution of the transverse positions. **c** The time evolution of transverse velocities. It is observed that the rising velocity oscillates with large amplitude in the zigzag stag when $t^* < 40$, however it is much more stable when the bubble enters the spiral stage

that the rising velocity fluctuates widely around $V^* = 1.3$ in the zigzag stage, while a sudden drop which leads the terminal velocity to $V^* = 1.2$ occurs when the bubble enters the spiral stage and the fluctuations are dampened. Such a change of the rising velocity when the bubble transits from the zigzag motion to a spiral one, is also reported by others [9, 10], who think it to be the effect of the dissipation in the wake.

However, what we mainly concern about is why the bubble motion transits from zigzag to spiral and the mechanism hidden behind the phenomena is not detected yet. Mougin and Magnaudet [9] obtained the spiral motion by imposing a mutual rotating disturbance on the zigzag path, therefore, they could not explain how the spiral stage was developed naturally, because they keep the bubble shape be frozen. We will further validate that the bubble deformation is very important in such a transition process. Shew and Pinton [10] suggested that a random perturbation would shift the double threaded wake to one side of the original zigzag plane, and the symmetry of the vortical force about the plane was broken. Then an induced torque would make the bubble rotate, which finally developed to a spiral motion. However, rather than validating this possibility, they thought that there might be another possible mechanism to induce such transition: two vortex lines of unequal strength would rotate about each other by mutual advection. They left the two possibilities to be an open question to unravel the causes of the transition from zigzag to spiral. It should be noted that with the help of the PIV measurement technique, Brücker [2] found that the double vortex threaded were not equal in strength when the bubble travels in a spiral path, however, for lack of further proof, this observation was not paid deserved attention.

As shown in the above section, the path instability of the rising bubble is closely related to the wake instability. Consequently, the time histories of the wake structures during the transition process from zigzag to spiral are present in Fig. 7.10, from $t^* = 40$ to $t^* = 50$. The dark color of the contour represents $\omega_z^* = -0.5$ while the light color is $\omega_z^* = 0.5$. Through a close inspection, it's observed that the new vortex pairs, which are just shedding from the bubble surface, gradually lose the symmetry structure from $t^* = 40$ to $t^* = 44$ and tend to rotate with each other. Later from $t^* = 45$ to $t^* = 49$, the two vortex threads finally twist with each other and thus a spiral motion of the bubble is induced. As a consequence, the asymmetry of the

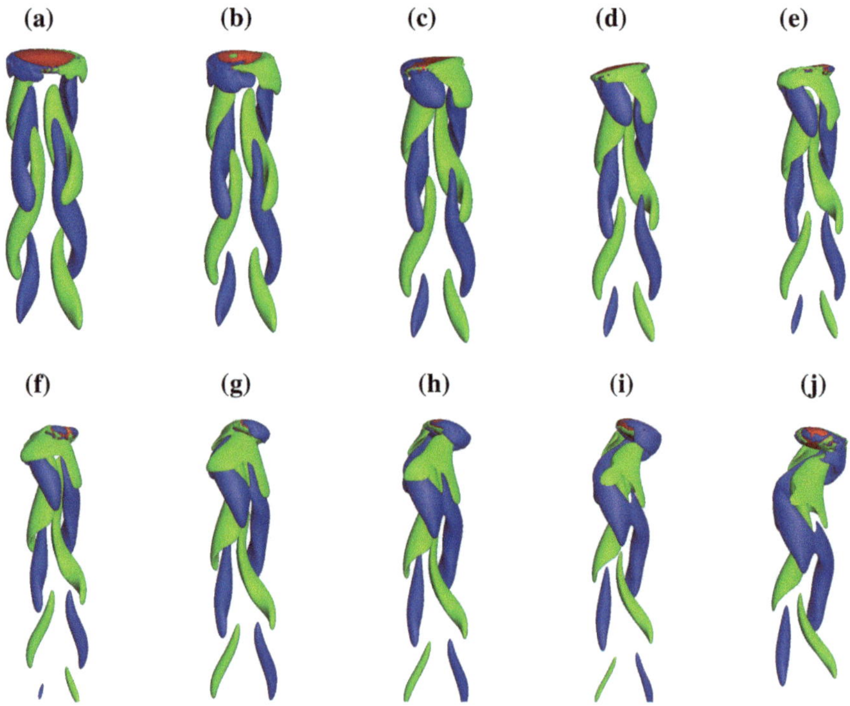

Fig. 7.10 The evolution of the vortex structures during the transition process from zigzag to spiral, within a time period of $t^* = 40 \sim 50$, the time interval is $\Delta t^* = 1$. The dark color is $\omega_z^* = -0.5$ while the light color is $\omega_z^* = 0.5$. It is observed that in this stage, the new shedding vortex pairs become asymmetric gradually, resulting in a rotation behavior between them

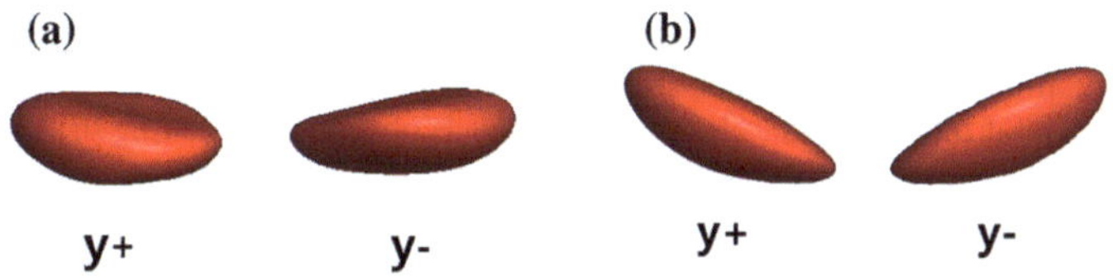

Fig. 7.11 The side views of the bubble shape when it travels in **a** spiral motion and **b** zigzag motion, respectively. It is observed that during the spiral motion, the bubble shape is no longer symmetrical, causing the shedding vortex pairs to be unequal in strength. By contrary, in the zigzag motion, the double vortex threads are equal in strength due to symmetrical bubble shape

vortex pairs seems to be responsible for the transition from zigzag to spiral, and we will try to verify this.

It is believed that the amounts of the vorticities accumulated on the bubble surface are directly depended on the curvature of the interface. Therefore, the side views of the bubble, where the negative ω_z^* and the positive ω_z^* are generated respectively, are present in Fig. 7.11. The side views of the bubble in the spiral motion are shown in Fig. 7.11a while the corresponding views of the zigzag motion are in Fig. 7.11b. It is obvious that in the spiral motion, the curvature at the y-side, where the positive vortices are shed from the bubble, is larger than that at the $y+$ side where negative vortices are shed. Consequently, the positive vorticities are gradually dominant in the vortex shedding process. In contrast, in the zigzag motion as shown by Fig. 7.11b, the two bubble sides where vortex pairs are shed, are exactly symmetrical, thus the strengths of the double vortex filaments are also equivalent.

To be more intensive on this issue, in case of $Bo = 2$, the distribution of the positive/negative streamwise vorticities on the bubble surface, as well as the interface curvature at corresponding times, are presented in Table 7.1. It should be noted that we calculate the maximum of the absolute value of the principal curvatures, referred as κ_{max} in the table. In the figures, we would find that the positive (negative) vortex thread is shed at the right (left) side of the bubble. Several time periods are selected: in the first stage from $t^* = 35$ to $t^* = 37$, the bubble still travels in the zigzag motion, it is obvious that positive and negative ω_z^* are symmetrically distributed at the two sides of the bubble, and correspondingly, κ_{max} are also equal in both vortex regions. However, in the second period when the bubble motion transits from zigzag to spiral, namely from $t^* = 41$ to $t^* = 47$, the positive vortices starts to prevail over the negative ones, at the meantime, the interface curvature also becomes larger at the positive vortex region. In the last stage after $t^* > 53$, when the spiral motion is finally generated, it is found that the positive vortices are much stronger than the negative ones, and the local interface curvatures is also more pronounced. Therefore, it is quantitatively validated that when the bubble shape is symmetrical at the two sides where positive and negative streamwise vortices are shed, the vorticity distribution on the bubble surface are also symmetrical, however, when the interface curvatures become asymmetric, the balance between the positive and negative vortices is also broken.

Table 7.1 (top row) The streamwise vorticity distributions on the bubble surface at different time periods; (bottom row) the contour maps of the maximum interface curvature at corresponding times. The value of the contour levels are attached. It is observed that during the zigzag period from $t^* = 35$ to $t^* = 37$, both the interface curvatures and positive/negative vorticity distributions are symmetrical, however, in the transition stage from $t^* = 41$ to $t^* = 47$, both symmetries are broken so that larger surface curvature results in stronger positive vortices, and after entering the spiral motion at $t^* > 53$, positive vortices completely dominate on the bubble surface because of the much larger local interface curvature

Time	$t^* = 35$	$t^* = 37$	$t^* = 41$	$t^* = 47$	$t^* = 53$	$t^* = 55$
ω_z^*						
κ_{max}						

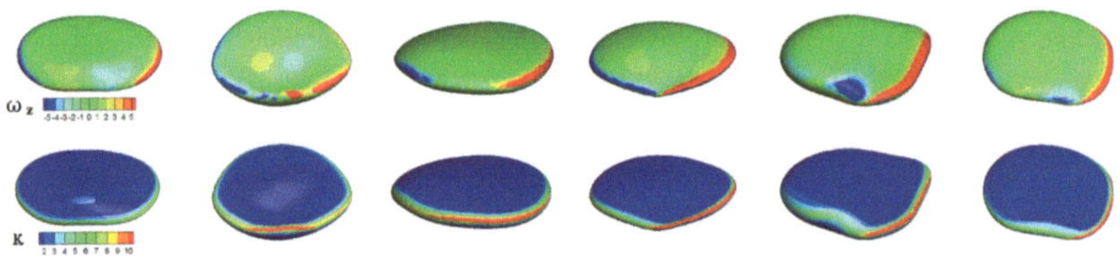

As interpreted in the previous section, the symmetric counter-rotating vortex pairs will induce a straight-line velocity, driving the bubble to rise in a zigzag manner. However, with unequal pairs, the result is much different that an angular velocity will be induced to drive the vortex pairs rotating around an invariant vorticity center [1] and sketched in Fig. 7.12a. When it comes to the current problem, the vortex pairs, as viewed from the top, are shown in Fig. 7.12b where the black outline is the bubble interface. It should be noted that the positive vortex filament in the innermost part is stronger than the negative one in the outer part. Then a counterclockwise angular velocity is induced to make the vortex pair rotating with each other. To conserve the momentum, the bubble will rotate in a clockwise direction so that a spiral motion is produced gradually. This conjecture is consistent with the numerical result shown in Fig. 7.1b, where a clockwise rotation of rising bubble is observed. Then a horizontal cross section, $4R$ downstream the bubble bottom, is cut out at $t^* = 50$. The isosurfaces of ω_z^* on the plane are present in Fig. 7.12c, where the contour values are given. As present, the structure of the counter-rotating vortices is much different from what we observe in the zigzag motion and the symmetry between the vortex pairs is broken. As shown in the figure, the inner part of the vortex pairs with positive sign dominate the direction of the induced vortex rotation because it is characterized by more maximum streamwise vorticity. To further validate that the inner part of the vortex pairs with positive sign is stronger than the negative one, the time history of the arithmetic sum of the streamwise vorticities, being $\tau_{plane} = \int \omega_z dS$ on the cut plane is present in Fig. 7.12d, which is calculated from $t^* = 70$ to $t^* = 79$ when spiral motion is completely formed. From the figure, τ_{plane} on the cut plane is always positive, indicating that the inner positive vortex thread must be dominant in the vortex pairs.

Therefore, it is the asymmetric vortex pairs, which are produced by asymmetric bubble deformation, to induce an angular velocity to drive the bubble to rotate in the opposite direction, and a spiral motion is finally formed. As a consequence, we answer the question raised by Shew and Pinton [11]: whether the spiral motion is induced by

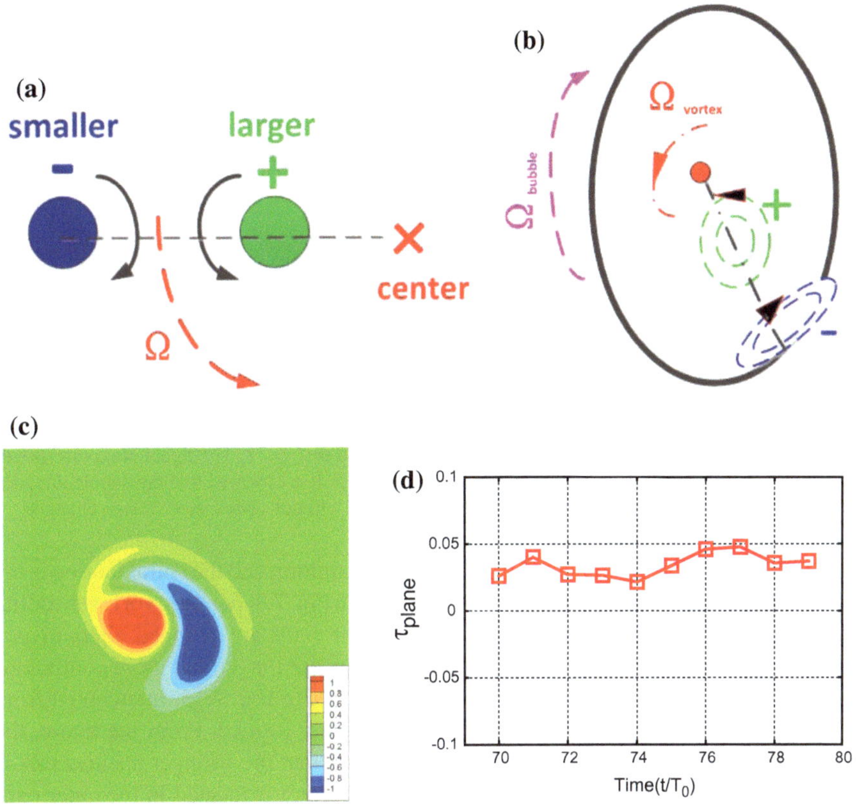

Fig. 7.12 **a** Sketch of the motion of the counter-rotating vortex pairs when they are with unequal strength. **b** The top view of the vortex structures in spiral motion, with $Ga = 50,\ Bo = 2$. **c** The contour of the vortex pairs in the cross plane $4R$ downstream from the bubble. **d** The time evolution of the arithmetic sum of $\tau_{plane} = \int \omega_z dS$ on the cross plane. As present, a counterclockwise angular velocity Ω_{vortex} is induced between the vortex pairs, and the bubble will therefore rotate in the clockwise direction with Ω_{bubble}. It is verified by the quantitative calculation that the arithmetic integration of ω_z^* on the bubble surface is always positive

the perturbation on the position of the vortex pairs or by the unequal-strength between the vortex pairs. This interpretation is consistent with the experimental observation by Brücker [2], who also suggested that the symmetrical vortex structure is broken when the bubble transits from the zigzag motion to the spiral motion. From Fig.12 in that paper, we find the same rule as we summarize that the bubble rotates in the opposite direction against the inner filament of the vortex pairs. However, he did not give any description and analysis about this, but it is actually the mechanisms to make the bubble rise in spiral trajectory.

Consequently, although there is no inclination for bubble to rotate in the clockwise or the counterclockwise direction when performing spiral motions, as observed in experimental studies, however, the relative strength between the double vortex threads actually determines the rotation direction.

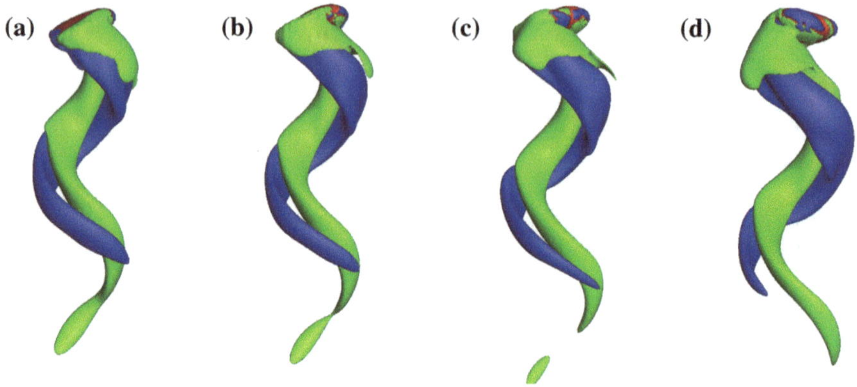

Fig. 7.13 The streamwise vortex structures behind the bubbles at different times during the spiral motion, respectively at **a** $t^* = 70$, **b** $t^* = 71$, **c** $t^* = 72$ and **d** $t^* = 73$. The light color corresponds to $\omega_z^* = 0.5$ while the dark color corresponds to $\omega_z^* = -0.5$. It is observed that the double threaded wakes tend to twist with each other and structure is stable without vortex shed or sign changed

Another distinct difference between the zigzag motion and the spiral motion is the structures of the double threaded wake. As shown in Fig. 7.4, periodic vortex shedding of ω_z^* is observed in the zigzag motion, and their signs are changed alternatively. However, in the spiral motion, the vortex filaments are longer and more continuous that the vortex shedding process vanishes, as shown in Fig. 7.13, intercepted from $t^* = 70$ to $t^* = 74$ when spiral motion is completely formed. From the figure, the double vortex filaments are twisted with each other that the stronger positive part is located in the inner side while the weaker negative part is located in the outer part. This observation is in very good agreement with the experimental results [4, 13] and the numerical results [9]. They all found that the vortex filaments no longer changed the sign in spiral motion, and the double threaded structure was very stable.

As concluded in Sect. 7.2.1, during the zigzag motion, the vorticities are evacuated from the bubble surface through periodic vortex shedding when a critical amount of ω_z^* is reached. However, in the spiral stage, the vortex pairs are more continuous without vortex shedding, a new way for vortex evacuation is developed that the vorticities will be evacuated from the bubble surface through the long filament with greater efficiency, correspondingly, the vorticities will not accumulate on the bubble surface any more. This is validated in Fig. 7.14a where the time evolution of the norm of ω_z^* accumulated on bubble surface is given, with $Bo = 2$. It is observed that in spiral motion when $t^* > 50$, the fluctuation is much smaller than that in the zigzag stage, indicating that the vortex evacuation is more stable. This stabilizing effect should be more conspicuous if the spiral path is more round. Thus the same calculation is carried out on the bubble of $Bo = 3$, whose trajectory is observed to be a pure spiral. The result presented in Fig. 7.14b shows that the amount of $\| \omega_z^* \|$ seems to be nearly constant when after the spiral motion starts.

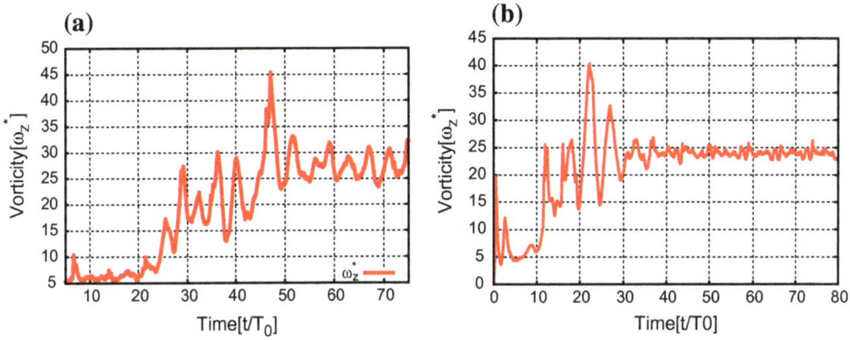

Fig. 7.14 **a** The amounts of the norm of ω_z^* accumulated on the bubble surface, with $Bo = 2$. **b** The amounts of the norm of ω_z^* accumulated on the bubble surface, with $Bo = 3$. It is observed that after the bubble enters the spiral stage, the evaluations of the vorticities from the bubble surface are more stable, and this stabilized effect is more conspicuous when the rising path approaches to a more spiral one

7.2.3 To Trigger the Spiral Motion by Imposing Perturbations on the Bubble Shape

In this part, our objective is to further prove that through imposing perturbations on the bubble interface, by which the symmetries of the bubble shape and the vortex pairs are broken, the bubble motion would transit from zigzag to spiral. This is viewed as a supplement of Sect. 7.2.2, in which the physical mechanics for the path transition is deeply discussed.

The flow condition under investigation is $\{Ga, Bo\} = \{35.36, 2.5\}$, in which the bubble motion is thought to be spiral according to the phase map given by [3]. However, we just observe this bubble to rise within a zigzag path, as presented in Fig. 7.15 where "*noperturbation*" is labeled. For comparison, still in Fig. 7.15 where "*perturbation*" is marked, small perturbations are imposed on the surface tension during the zigzag motion at $t^* = 62$ for a short time period of $\Delta t^* = 0.5$, then the asymmetric bubble shape will be triggered, and as a result, the double-threaded vortices will lose symmetry gradually. In the left column of the figure, the projection of the bubble rising paths are displayed with and without perturbations respectively, whereas the red (green) lines indicates the paths before (after) imposing the perturbations. In other right columns, the evolution of the streamwise vorticity iso-contour $\omega_z^* = \pm 0.4$ are presented, and meanwhile, the bubble shapes at the corresponding time points are also attached that the distribution of ω_z^* on the bubble interface are plot.

Without imposing perturbations, it is observed that both the bubble shape and the vortex pairs keep symmetric structures throughout the computations, as a consequence, the bubble rises within a zigzag motion as we observe. The computation is stopped at $t^* = 82$ due to the limitation of the computational resources. In contrast, after imposing perturbations on the surface tension during the numerical simulations,

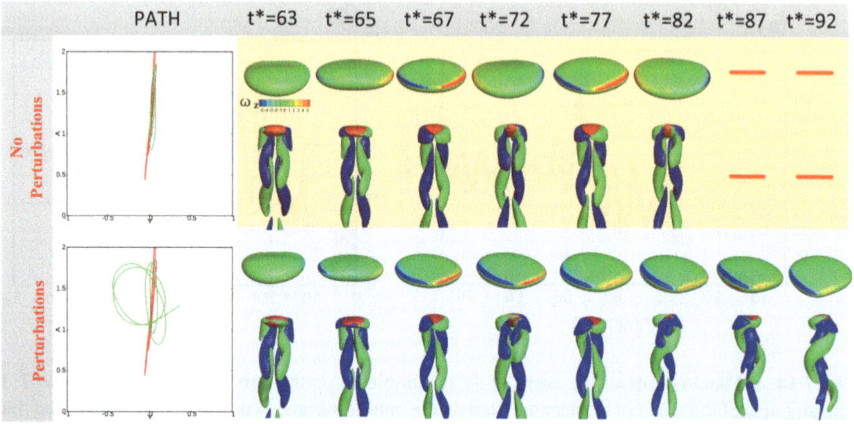

Fig. 7.15 The characteristic properties of the bubble motion without and with imposing perturbations on the surface tension. From left to right: the projection of the rising path of the bubble and the evolution of the streamwise vorticity iso-contours $\omega_z^* = \pm 0.4$ at different time periods, in addition, the streamwise vorticity distributions on the bubble interface at corresponding times are attached. It is observed without perturbations, the bubble rises within a zigzag path and the flow characteristics are symmetric, in contrast, such symmetries are broken when imposing the perturbations, and the bubble motion transits from zigzag to spiral gradually

the bubble motion transits from zigzag to spiral gradually. To explore the physical mechanics of such transition, we should still look into the evolutions of the bubble shape and vortex structures during this transition stage. It is observed that from $t^* = 63$ to $t^* = 72$, the bubble shape and the distribution of ω^* on the interface develops to be asymmetric gradually, and the shedding double-threaded vortices tend to twist with one another. Subsequently, from $t^* = 77$ to $t^* = 92$ when the spiral motion is finally generated, the negative vorticities distributed on the bubble interface are dominant over the positive ones, and the shedding vortex pairs twist with one another completely. The evolution process of vorticity distribution during the transition stage is identical with Table 7.1, and the only difference between them is which thread of the vortex pairs is stronger: positive one induces anticlockwise spiral motion while negative one in duces clockwise motion.

Therefore, this part of the work contributes to strengthen the interpretations we give in Sect. 7.2.2, it shed more lights on the relations between the bubble shapes, the vortex structures and the bubble motions.

7.3 Magnetic Influence of the Spiral Motion

After detecting the physical mechanism for path transition from zigzag to spiral, we intend to validate that by diminishing the unbalance between the asymmetric vortex pairs, the radius of the spiral motion will be decreased. As a consequence, the vertical MF would be a good choice because the streamwise vorticities would be greatly weakened under the influence of vertical MF, this is validated in our previous study in Chap. 5. Therefore, we will also apply vertical MF on the spirally rising bubble, to make our suppositions be fully confirmed. We select the rising bubble within $Ga = 50$ and $Bo = 3$ as the reference case, which performs a pure spiral motion when no magnetic field is imposed, as shown in Fig. 7.1c. Initially, the bubble is rising freely without magnetic field till $t^* = 42$, whereas the spiral motion has been generated already. At the moment of $t^* = 42$, different vertical magnetic fields are imposed on the spiral bubble suddenly, respectively with $N = 0.3, N = 0.5$ and $N = 1.0$, in order to examine the variations of the the bubble paths as well as the vortices under different magnetic strengths.

Firstly, we will show the variation of the rising velocity under different magnetic strength. In Chap. 5, it is presented that before the bubble rising rectilinearly, the rising velocity of the bubble will be increased due to the magnetic effect. This is again validated in Fig. 7.16, which shows different rising velocities when N is varied. From the figure, it is observed that except for the relatively lower rising velocity under $N = 0$, there is almost no difference among other cases under $N = 0.3, N = 0.5$ and $N = 1.0$. Therefore, we do not think the variation of the rising velocities will play a notable role if the bubble path changes significantly.

After that, we present the arithmetic sum of ω_z integrated on the bubble surface before and after imposing the magnetic field, with $\tau_{bubble} = \int_S \omega_z dS$ where S is the interface of the bubble. The results are plotted on Fig. 7.17, whereas the red (green) lines depict the amount of Ω before (after) $t^* = 42$. From Fig. 7.17a, it is observed that without magnetic field, τ_{bubble} is almost stabilised at $\tau_{bubble} = 3$, indicating that there are more positive vorticities accumulated on the bubble interface. However, in Fig. 7.17b where the magnetic field is imposed with $N = 0.3$ after $t^* = 42$, the value of τ_{bubble} is found to decrease to $\tau_{bubble} = 2$. Furthermore, by increasing the magnetic strength continually to $N = 0.5$ and $N = 1$ after $t^* = 42$ respectively, the value of τ_{bubble} further drops to $\tau_{bubble} = 1.5$ and $\tau_{bubble} = 0$ correspondingly. Therefore, from the figures, it is observed that under the influence of vertical magnetic fields, the

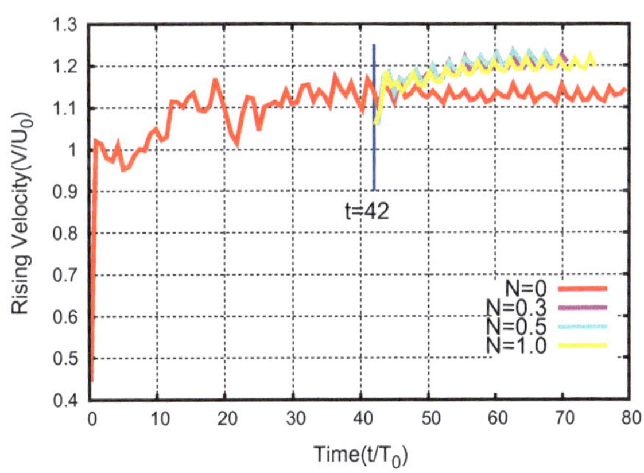

Fig. 7.16 Time histories of the rising velocities versus time under different magnetic field. It is observed that except for the relatively lower velocity under $N = 0$, there is not much difference among other three cases

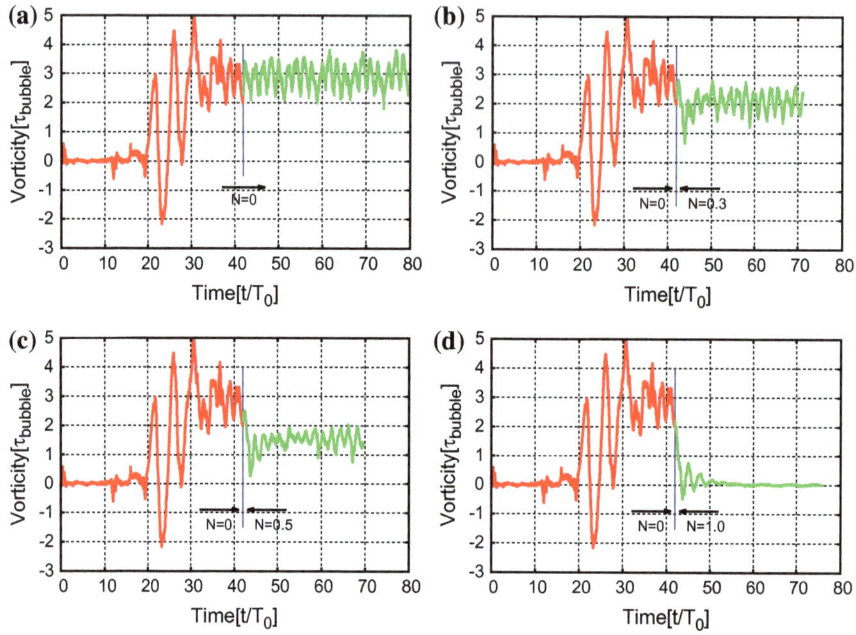

Fig. 7.17 Time histories of the arithmetic sum of ω_z integrated on the bubble surface before and after imposing a vertical magnetic field at $t^* = 42$, from left to right: **a** $N = 0$, **b** $N = 0.3$, **c** $N = 0.5$, **d** $N = 1.0$. It is found that as the magnetic strength is increased to $N = 0.3$, $N = 0.5$ and $N = 1.0$, the arithmetic sum of $\tau_{bubble} = \int_S \omega_z dS$ is decreased from $\tau_{bubble} = 3$ to $\tau_{bubble} = 2$, $\tau_{bubble} = 1.5$ and $\tau_{bubble} = 0$ respectively. The results indicate that on the bubble interface, the asymmetry between the amounts of the positive and negative vortices are diminished under stronger magnetic field

amounts of the positive and negative vortices are more symmetric on the bubble surface.

As a consequence, we have reasons to believe that the radius of the spiral motion will be reduced under a strong magnetic field, because the asymmetry between the vortex pairs are decreased. Under different magnetic fields, the top views of the spiral motions are present in Fig. 7.18, whereas red (green) lines are for paths before (after) $t^* = 42$. From the figures, it is clearly shown that after imposing the magnetic field at $t^* = 42$, the radius of the spiral motion is decreased significantly, complying with $r_{N=0} > r_{N=0.3} > r_{N=0.5}$. This variation trend is ascribed to the more balanced counter-rotating vortex pairs. Moreover, when the magnetic field is as strong as $N = 1.0$, the radius of the spiral motion is deceased to zero, that is because the motive power for spiral motion is disappeared as $\tau_{bubble} = 0$. Therefore, we confirm that it is the asymmetric vortex pairs to provide the power for spiral motion, the radius of which can be reduced by decreasing the vortex asymmetry. Furthermore, by comparing Fig. 7.17d with Fig. 7.18d, it is observed that even τ_{bubble} has decreased to zero at $t^* = 50$ already, the bubble still travels within a weakened spiral motion.

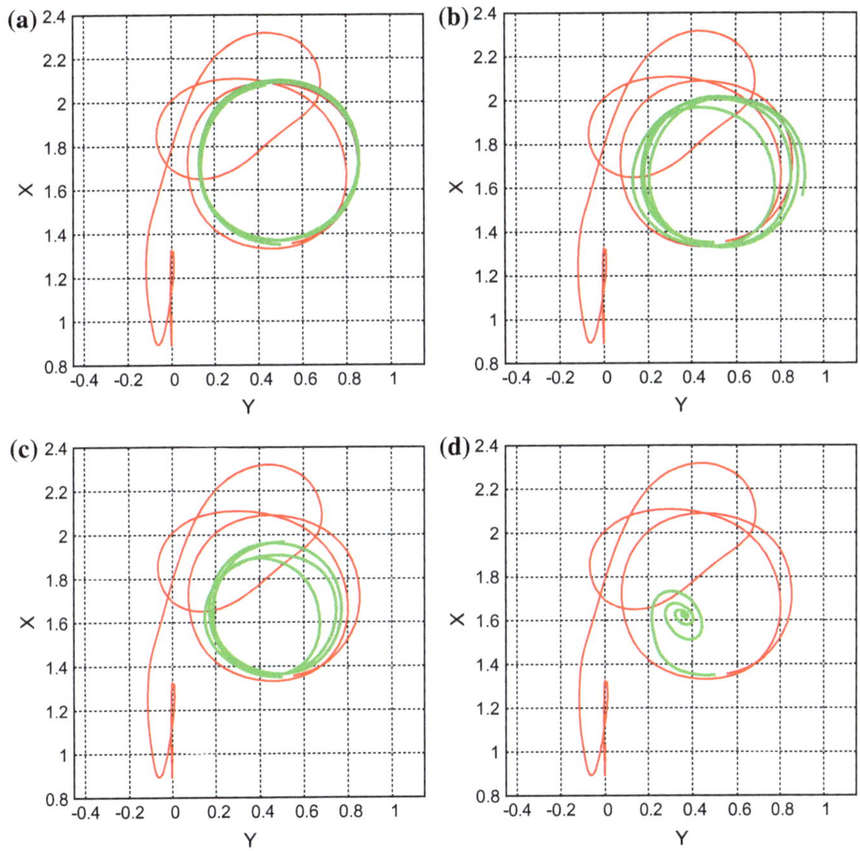

Fig. 7.18 The evolution of the spiral motion of the bubble before and after imposing a vertical magnetic field at $t^* = 42$, from left to right: **a** $N = 0$, **b** $N = 0.3$, **c** $N = 0.5$, **d** $N = 1.0$. It is found that when the magnetic field is stronger, the radius of the spiral motion is decreased, because the strength of the double vortex threads are more symmetrical as shown in Fig. 7.17. Particularly, under $N = 1.0$, the spiral motion seems to be disappeared completely because of $\tau_{bubble} = 0$ in such case

Therefore, it indicates that the variation of the spiral path falls behind the variation of τ_{bubble}, and it is the vorticity to determine the rising path.

7.4 Summary

This chapter has detailed a mechanism explaining why and how the bubble travels from zigzag to spiral. Regarding rising within the zigzag motion, we find it is the accumulation of the streamwise vorticities on the bubble surface, rather than the total vorticities, that trigger the vortex shedding in a zigzag period. When the bubble transits from zigzag to spiral, previous studies always described the different behaviors of the bubble in the two travelling patterns, or different vortex structures behind the bubble. Our mechanism, which is based on the quantitative calculation of the counter-rotating vortex strengths, however, explains how different vortex structures are evolved, and how an angular velocity is induced between the asymmetrical vortex pairs to drive the bubble traveling from zigzag to spiral. Furthermore, by investigating the evolution of the interface curvature and vortex distribution on the bubble interface during the zigzag/spiral transition stage, we find the asymmetric deforming of the bubble shape causes the unbalanced vortex distribution, where more vorticities are shed from the interface with larger local curvature. To further validate this, in the spiral rising regime, we are able to produce asymmetric bubble shape and vortex shedding by imposing perturbations on the surface tension of the bubble, and a path transition from zigzag to spiral is correspondingly induced.

In the second part of the paper, by imposing external vertical MFs, which are used to diminish the asymmetry between the vortex pairs, the spiral motion is observed to be weakened that finally transits into the rectilinear path under much stronger MFs. Nevertheless, we give detailed explanations about why not a zigzag motion is recovered, it is because not only the symmetric vortex pairs are needed, but also strong enough vortex pairs are required to trigger vortex shedding. The influences of vertical MFs on vortex structures are also reported in detail in Chap. 5, and the present study is rather a particular case, which only focuses on the spiral stage of the bubble.

Furthermore, to make things simple, we only investigate the bubble motion in low *Re* flows, in which almost no other flow instability happens, therefore, the imbalance between the vortex pairs can be ascribed to the asymmetric deforming of the bubble shape during the motion. However, when *Re* becomes higher, other instabilities in the flow field would also lead the vortex pairs to be asymmetric, and thus spiral or even chaos rising trajectories are observed.

References

1. Batchelor GK (2000) An introduction to fluid dynamics. Cambridge University Press, Cambridge
2. Brücker C (1999) Structure and dynamics of the wake of bubbles and its relevance for bubble interaction. Phys Fluids 11(7):1781–1796
3. Cano-Lozano JC, Martinez-Bazan C, Magnaudet J et al (2016) Paths and wakes of deformable nearly spheroidal rising bubbles close to the transition to path instability. Phys Rev Fluids 1(5):053604
4. De Vries AWG, Biesheuvel A, Van Wijngaarden L (2002) Notes on the path and wake of a gas bubble rising in pure water. Int J Multiph Flow 28(11):1823–1835
5. Ellingsen K, Risso F (2001) On the rise of an ellipsoidal bubble in water: oscillatory paths and liquid-induced velocity. J Fluid Mech 440:235–268
6. Lunde K, Perkins RJ (1997) Observations on wakes behind spheroidal bubbles and particles. Vancouver, ASME fluids engineering division summer meeting, Canada, p 141
7. Lunde K, Perkins RJ (1998) Shape oscillations of rising bubbles. fascination of fluid dynamics. Springer, Dordrecht, pp 387–408

8. Magnaudet J, Mougin G (2007) Wake instability of a fixed spheroidal bubble. J Fluid Mech 572:311–337
9. Mougin G, Magnaudet J (2001) Path instability of a rising bubble. Phys Rev Lett 88(1):014502
10. Shew WL, Pinton JF (2006) Dynamical model of bubble path instability. Phys Rev Lett 97(14):144508
11. Shew WL, Poncet S, Pinton JF (2005) Path instability and wake of a rising bubble. preprint, ENS, Lyon
12. Tripathi MK, Sahu KC, Govindarajan R (2015) Dynamics of an initially spherical bubble rising in quiescent liquid. Nat Commun 6:6268
13. Veldhuis C, Biesheuvel A, Van Wijngaarden L (2008) Shape oscillations on bubbles rising in clean and in tap water. Phys Fluids 20(4):040705
14. Zenit R, Magnaudet J (2008) Path instability of rising spheroidal air bubbles: a shape-controlled process. Phys Fluids 20(6):061702

Chapter 8
Conclusions and Prospect

8.1 Conclusions

In order to improve the performance of the numerical methodologies in studying the incompressible MHD flows, a new solver is developed and implemented into the open source Gerris code. The new Gerris-MHD solver, which is based on the Cartesian grids and the AMR technique, is able to simulate the single-phase MHD flows respectively of being bounded by electrically insulating or conducting walls with a cut-cell approach. In addition, by using the VOF method, the solver is also able to solve the multi-phase MHD flows and the discontinuities of the physical properties across the fluid-fluid interface can be handled very well. After that, the single bubble motion with or without the influence of the MF is investigated numerically, and the physical mechanisms are discussed in details. The main contents and conclusions are listed in the following:

1. The cut-cell method belonging to the Cartesian grids technique is introduced to simulate the MHD flows bounded by the electrically insulating boundaries of complex geometries. It liberates the numerical modeler from the laborious work of generating unstructured or body-fitted grids, which are traditionally used to simulate such type of flows. Besides, by introducing the AMR technique, the mesh sizes could be varied with several orders of magnitude from the boundary layer to the center of the flow field, and therefore, the computational efficiency is greatly improved by preserving the numerical accuracy. Moreover, the key issue in designing an accurate numerical scheme based on the AMR framework is to keep the consistency of the flux calculation on the coarse/fine cell faces, and we prove without such a treatment, the convergence of the numerical results will decrease to first-order.
2. When the complex solid walls are electrically conducting, then a VOF method is used to estimate the partial volumes of the fluid and the solid in a "mixed-cell", as well as the partial areas on a "mixed-face", and then by combining it with the cut-cell approach, the electromagnetic field in the domain is solved as a whole

J. Zhang, *The Developments and the Applications of the Numerical Algorithms in Simulating the Incompressible Magnetohydrodynamics with Complex Boundaries and Free Surfaces*, Springer Theses, https://doi.org/10.1007/978-981-10-6340-4_8

on the unified grids. This approach is typically different from the traditional numerical method based on two set of grids separating the fluid and the solid, and it is more efficient by leaving out the information transmission across the solid-fluid boundary.

3. With the aid of accurate numerical methods in modelling the surface tension force within Gerris, such that by using the CSF model to enhance the numerical stability, by using the balanced-force technique to reduce the spurious flow and by using the height function to estimate the curvature, the new Gerris-MHD solver is able to simulate the MHD flows with free surface, even if the electrical conductivity in the liquid metal is ten thousand of times higher than that in the gas. In addition, a particular free surface flow with Marangoni stresses induced by the temperature gradient is also implemented in the solver, and accurate results are obtained by performing a series of numerical tests.
4. The dynamics behaviors of a single argon bubble rising in the liquid GaInSn under the influence of vertical MFs are investigated numerically, and the results remedy partially some flaws of the experimental studies, particularly the limited measurements of the flow structures in the experiments because of the opacity of the liquid metal. Nevertheless, not only the nonmonotonous influence of the vertical MF on the bubble motion is also observed in the numerical studies, that is moderate MF promotes the bubble motion while strong MF suppresses it, we also have the best opportunity to investigate the physical mechanisms responsible for this. It is because the bubble shapes to be more flat under the influence of the vertical MFs, and therefore the vortex threads are more parallel with one other to cause more rectilinear rising paths, to this extent, the bubble motion is accelerated by the vertical MF. After the bubble rises in a fully rectilinearly path, stronger vertical MFs tend to block the bubble motion because the pressure difference at the top and the bottom of the bubble is enlarged due to the Lorentz force. As a consequence, the tangled web of relationships between the bubble shapes, the bubble wakes and the rising paths under the influence of vertical MFs is more clear.
5. The similar numerical investigations of a single argon bubble rising in the liquid GaInSn under the influence of horizontal MFs are further conducted, and the most striking observations are the anisotropic phenomena in the flow field, such as the bubble shape, the vortex structures around the bubble and the double-threaded wakes behind the bubble. More than that, the physical mechanisms to cause such anisotropic characteristics are discussed by computing the distributions of the Lorentz forces in vicinity of the bubble. It is because the inhomogeneous Lorentz force produces an induced vortex to elevate the pressure at the bubble waist in the field direction, while the pressure at the waist perpendicular to the field direction is almost unchanged, and such influence on the pressure distribution finally leads to the anisotropic effect in the flow field. This observation is also consistent with some theoretical predictions, which believe that the flow field should be more homogeneous in direction parallel to the MF.
6. After understanding the close relation between the vortex structures and the rising paths of a bubble, I return to one of the main problems in the bubble

dynamics: why a single bubble sometimes transits from zigzag to spiral? By carrying out a series of numerical simulations, we confirm that for a zigzag rising bubble, vortex pairs are shed twice during a period of the pure zigzag path and the vortex pairs are in equal-strength under such circumstance. Nevertheless, if the balance between the counter-rotating vortices is broken, an angular velocity is induced between the asymmetric vortex pairs to the bubble to rise in an opposite spiral path. Furthermore, the imbalance between the vortex pairs are caused by the asymmetric deformation of the bubble by detailed investigations in the low Re flows. Moreover, this assumption is further validated by imposing external vertical MFs to weaken the imbalance between the vortex structures, and correspondingly, the rotating radius of the spirally rising bubble is observed to decrease.

8.2 Prospective

On basis of the numerical algorithms proposed in this thesis, new implementations are still under development to extend the application range of the Gerris-MHD solver in simulating more complex multi-phase MHD flows, whereas the liquid-vapor phase change is involved. However, this is more challenging because the jump conditions across the interface are more severe and the muti-physical fields are coupled more closely. Besides, some other new scopes of MHD multi-phase flows are also under investigation based on the solver, such as the thermal migration of the droplet under the influence of MF, the motion of the droplet across the inhomogeneous MF, and the MHD effect on the droplet impacting problems onto liquid or solid surfaces. They are widely encountered in the fusion devices and the metallurgic engineering, however, due to the limitations of the measuring techniques in the experiments, the quantitative investigations are difficult to achieve so far. Our numerical simulations provide an alternative perspective.

Regarding the bubble motion under the influence of the MFs, although the mechanisms in case of a single bubble are very clear, however, the applications in engineering area mainly concern to the bubbly driven flows that hundreds of bubbles will interact with each other, and hence the flows are more complex. As a consequence, the magnetic influence on the interactive behaviors between the bubbles need to be investigated firstly, and after that, a suitable turbulent model or accurate DNS should be tested to simulate such bubbly flows. Quite evidently, there is still a long way to go in the future before solving these problems completely.

MIX
Papier aus verantwortungsvollen Quellen
Paper from responsible sources
FSC® C105338

If you have any concerns about our products,
you can contact us on
ProductSafety@springernature.com

In case Publisher is established outside the EU,
the EU authorized representative is:
Springer Nature Customer Service Center GmbH
Europaplatz 3, 69115 Heidelberg, Germany

Printed by Libri Plureos GmbH
in Hamburg, Germany